IT Solutions for the Smart Grid

Tobias Brandt

IT Solutions for the Smart Grid

Theory, Application, and Economic Assessment

 Springer Vieweg

Tobias Brandt
Freiburg, Germany

Dissertation, University of Freiburg, Germany, 2015

ISBN 978-3-658-12414-4   ISBN 978-3-658-12415-1 (eBook)
DOI 10.1007/978-3-658-12415-1

Library of Congress Control Number: 2015960462

Springer Vieweg

Printed on acid-free paper

Springer Vieweg is a brand of Springer Fachmedien Wiesbaden
Springer Fachmedien Wiesbaden is part of Springer Science+Business Media
(www.springer.com)

Acknowledgements

Over the past four years, I have had the fortune to work as part of an exquisite team that has enabled me to write the dissertation at hand. First and foremost, I want to thank my academic advisor and mentor, Prof. Dr. Dirk Neumann, for guiding, supporting, and challenging me during these years. I thank my partners in crime–Johannes, Sebastian, and Stefan–for being the best colleagues one could wish for. Your knowledge and your skills have enhanced my work substantially and I can only hope that I have been able to reciprocate this gift. Furthermore, I would like to thank Carla Li-Sai for her care and support, protecting us from the day-to-day madness of university bureaucracy and just making everybody feel welcome at the office.

Throughout my time as a doctoral student, I have met, worked with, and learned from some extraordinary people and researchers. I thank Dr. Michael Stadler for hosting me as a visiting researcher at Lawrence Berkeley National Laboratory in 2014. This semester in California was a truly inspiring time and I am grateful to everybody who helped make this an experience I will cherish for the rest of my life. I also want to extend my gratitude to Prof. Dr. Dr.h.c. Günter Müller for the valuable feedback he provided during our doctoral seminars, as well as to Dr. Sebastian Busse and Dr. Fabian Lang for successful collaborations and great friendships. I would not have had this freedom, time, and independence to conduct my research without a Ph.D. scholarship from the Foundation of German Business (Stiftung der Deutschen Wirtschaft). I am very grateful for the opportunity

I was given and the inspiring people I met during our seminars and workshops.

My very special thanks go to my family–my parents, Karin and Horst, and my siblings, Yvonne and Hendrik–whose love and support have been a reliable certainty throughout my life; and to my girlfriend Julia who has given me comfort when I needed it, kept my from being overly dramatic when I needed it, and told me to get done with this thing when I needed it the most.

Tobias Brandt

Contents

I Introduction: Integrating Sustainable Energy Technologies **1**
1   Motivation: A Changing Energy Landscape   1
1.1 Lifeblood of Society   2
1.2 Adapting to a Changing World   3
1.3 The Transition towards Renewable Energy   5
2   Smart Grids   6
2.1 The Smart Grid Architectural Model   7
2.2 Energy Informatics   10
3   Research Framework   12
4   Thesis Summary   16
4.1 Summary: Designing IT Solutions for Individual Households   16
4.2 Summary: Designing an Energy Information System for
     Microgrid Operation   18
4.3 Summary: A Business Model for Employing Electric
     Vehicles for Energy Storage   19
4.4 Summary: Mechanism Stability—Flash Crashes and Avalanche Effects   20
5   Conclusion   21
5.1 Key Insights   22
5.2 Outlook   25
6   References   27

II Designing IT Solutions for Individual Households . . .   **33**
1   Introduction   34

2   Related Work   35
2.1 Research Domain: Energy Informatics   35
2.2 Research Approach: Design Science   37
2.3 Research Gap: IS Design in Smart Grids   37
3   Research Model   39
3.1 Definitions   40
3.2 Modeling IS Design in the Smart Grid   42
4   Case Study: Green Synergies   44
4.1 Setting and Motivation   44
4.2 Information System Design   46
4.3 Testing and Evaluation   49
5   From Objectives to Outcomes: Preventing Interferences .   55
5.1 From Design Objectives to the IT Artifact   55
5.2 From the IT Artifact to the Technical Object   57
5.3 From Enhancing Features to Enhancing Affordances . .   59
6   Conclusion   60
7   References   62
8   Appendix   68
8.1 Physical System   68
8.2 Implemented Strategies   71

III Designing an Energy Information System for Microgrid
     Operation **73**
1   Introduction   74
2   Relevance to IS Research and Related Work   76
2.1 Green IS and Energy Informatics   76
2.2 General Research on Microgrids   78
3   Problem Statement and Research Design   80
4   Information System Design and Evaluation   83
4.1 Forecasting Modules   84
4.2 Optimization Module   88
4.3 Evaluation Setting   91
4.4 Evaluation Results   93
5   Discussion   99
5.1 Technical and Economics Implications   99

5.2 Implications for IS Research 101
6   Conclusion 103
7   References 104
8   Appendix 108

IV A Business Model for Employing Electric Vehicles for
Energy Storage **111**
1   Introduction 112
2   Research Framework and Related Work 114
2.1 Electric Mobility and Renewable Energy 115
2.2 Electric Vehicle Aggregation and Frequency Regulation . 116
2.3 Parking Garage Operators as Intermediaries 117
2.4 Business Models 118
3   Modelling the Business Case 119
3.1 Product 119
3.2 Customer Interface 122
3.3 Infrastructure Management 123
3.4 Financial Aspects and Business Model Synopsis 124
4   Evaluation of the Business Model 126
4.1 Market for Frequency Regulation 126
4.2 Data Sources 130
4.3 Modeling Annual Revenue 133
4.4 Results 135
5   Discussion 138
6   Conclusion 140
7   References 141

V Mechanism Stability—Flash Crashes and Avalanche Ef-
fects **149**
1   Introduction 150
2   Related Work 152
2.1 Market Crashes 152
2.2 Information Processing and Competitive Advantage . . . 154
2.3 Game Form and Hypergame Theory 155
2.4 Contribution 156

3 The Role of Information in Game Form Determination . 156

3.1 Sequential and Simultaneous Games 157

3.2 The Influence of Information Technology on Game Form 160

4 The Impact of Strategic Misperceptions 162

4.1 The Hypergame for Two Players 162

4.2 Misperceptions in Games with Many Players 166

4.3 Numerical Examples 170

5 Implications, Interaction with Herding, and Limitations . 173

6 Application to Demand Side Management 176

7 Conclusion 179

8 References 180

9 Appendix 184

List of Figures

I–1   Smart Grid Architectural Model (adapted from CEN-
      CENELEC-ETSISmart Grid Coordination Group, 2012)   8
I–2   Energy Informatics Framework (as in Watson et al.,
      2010)   10
I–3   Contribution of thesis to a smart grid implementation   12
I–4   Structure of thesis according to scale of applications   16

II–1 Starting point: IS design for the Smart Grid   39
II–2 Design-Interference Model for the Smart Grid   43
II–3 Visualization of the information system functionality
      according to constructs of the Design-Interference Model   48
II–4 Functionality of the IT artifact   48
II–5 Composition of energy supply to household for differ-
      ent strategies   54
II–6 Design-Interference Model with guidelines to prevent
      interferences   61
II–7 Schematic representation of physical system   70

III–1 The microgrid setting in the Energy Informatics Frame-
      work, based on Watson et al. (2010)   77
III–2 Electrical network at the test site   80
III–3 Research design of the microgrid project   81
III–4 Schematic representation of the information system .   84
III–5 FFT-based load forecaster with time in 15-minutes
      intervals and load in kW   86

III–6 Actual PV efficiency for Jan 29 and Jan 30, 2013.
Light gray columns indicate reported clear sky. . . .   95

IV–1 Research framework   114
IV–2 Market position of parking garage operator (interme-
diary)   120
IV–3 Business model for EV4ES by a parking garage operator 125
IV–4 Daily auctions for tertiary control reserve on the GCRM 128
IV–5 Prices at the GCRM for tertiary reserves in 2014 . .   131
IV–6 Data on electric vehicles   132
IV–7 Occupancy of two sample parking garages in Freiburg   133
IV–8 Average number of vehicles available and maximum
amount of negative regulation power supplied (Simu-
lation with 10,000 vehicles per day)   136

V–1 Informational edge and game form determination . .   156
V–2 Sequential and simultaneous game   158
V–3 Subjective game G^i   164
V–4 Sequential and simultaneous game   165
V–5 Numerical example of two-player game   170
V–6 Numerical example of three-player game   171
V–7 Payoffs in 25-player game with increasing number of
misperceptions   172
V–8 Price movements in a sequential game (a), a simultan-
eous game with misperceptions (b), a simultaneous
game with herding (c), and a simultaneous game with
misperceptions and herding (d)   175

List of Tables

I–1   Allocation of research papers within SGAM   15

II–1 Evaluation parameters   51
II–2 Annual energy costs and annual effect on the grid . .   52
II–3 Constraints on physical system   69
II–4 Constraints for benchmark (B)   72
II–5 Constraints for decision strategies (S1 and S2) . . .   72

III–1 Variables and parameters in the optimization problem   89
III–2 Comparison of forecasted and actual loads (all values
      in percent)   93
III–3 Comparison of forecasted and actual PV generation
      (per MWp installed) for each model   96
III–4 Results of system evaluation for 3 MWp installed . .   98
III–5 Results of system evaluation for 5 MWp installed . .   99

IV–1 Revenue evaluation (all values in EUR)   137

1 Introduction: Integrating Sustainable Energy Technologies

1 Motivation: A Changing Energy Landscape

For decades, global energy consumption has only known one direction—upwards. Energy fuels cars, trucks, and planes, it powers machines, it allows to cook, helps to communicate, and enables to work. Energy is the foundation of prosperity in developed countries and the engine of growth in the developing world. Yet, despite its importance, popular awareness and discussions on how we produce, distribute, and consume energy used to be rare. With the turn of the century, the public perception has changed. Solving the *Energy Crisis* has suddenly become a grand challenge of the 21St century (Armaroli & Balzani, 2007). Germany has proclaimed the *Energiewende* (energy turnaround) and Silicon Valley has become a harbor for clean energy startups. Europe, the United States, China, and other countries around the globe are searching for ways to transition beyond the fossil-fuel-based energy paradigm of the 20th century. While energy has always been everywhere, talk of it is now, as well.

In this section, this shift in perception is investigated by, first, describing the relevance of energy to societies and economies around the globe. Second, reasons for the unsustainability of the traditional energy paradigm are explained. Third, the challenges that a shift to a sustainable and renewable energy supply is facing and how they relate to the *Smart Grid* concept investigated in this thesis are outlined.

1.1 Lifeblood of Society

The term *energy* is commonly used with a meaning different from its strict physical definition. While many physics textbooks define energy as the ability of a system to perform work, Doménech et al. argue that "energy may be conceived, in a first approximation, as the capacity to produce transformations" (Doménech et al., 2007, p. 51). They also emphasize that energy is one of the most abstract and difficult to grasp concepts in physics. The popular notion of energy is largely associated with electricity, particularly electric current (Tarciso Borges & Gilbert, 1999). More broadly, energy is considered to be what makes cars, laptops, and coffee machines run—which is not that different from Doménech et al.'s definition. For the most part of this thesis, *energy* is used with this colloquial meaning, as in *energy industry* or *energy systems*. Thus, the latter includes electrical power grids and circuits, but also systems transferring and storing other forms of energy (e.g. the potential energy stored in gas or oil). Furthermore, within these general discussions, *energy* and *power* are used interchangeably. However, the thesis contains a number of equations describing particular energy systems. Naturally, with respect to these equations, the terms *energy* and *power* follow their respective physical definitions.

Regardless of the definition used, the evolution of humans and human society is closely tied to the access, transformation, and use of energy. Burton (2009) summarizes how the control of fire affected human physiology and allowed mankind to evolve beyond its primate relatives. Wrangham further emphasizes this point, arguing that the rise of humanity and the foundations of human societies "stemmed from the control of fire and the advent of cooked meals" (Wrangham, 2010, p. 2). White suggests that "man would have remained on the level of savagery indefinitely if he had not learned to augment the amount of energy under his control and at his disposal for culture-building by harnessing new sources of energy" (White, 1943, pp. 340-341). Pimentel & Pimentel (2008) outline how this access to new sources of energy laid the foundation of civilization and modern prosperity. The necessary amount of human labor in agriculture was first reduced by draft-animals and later by windmills and waterwheels. Each step enabled people to invest more time in other ventures, increasing the division of labor and the complexity of social systems. This development culminated in the Industrial Revolution through the invention of the steam engine, "[signaling] the beginning of the use of fossil fuels as an energy source" (Pimentel & Pimentel, 2008, p. 2).

This first industrial revolution was followed by a second, which brought mass production powered by electrical energy (Chandler, 1992). Since then, electronics, digitalization, the Internet, and wireless communication have been and still are transforming industries, economies, and societies. The *Internet of Things* (Höller et al., 2014) and *Industry 4.0*[1] enable communication and coordination between devices and machines, outlining the possibility of further gains in efficiency and productivity.

The intimate relationship between oil and wars exemplifies that the consequences of the strive for new energy sources were not exclusively positive (Heinberg, 2005; Le Billon & El Khatib, 2004; Ross, 2006). Nevertheless, the dependence of human prosperity on energy resources is fundamental. Hence, a shift in the accessibility of energy resources has a profound impact on societies, politics, and people around the globe.

1.2 Adapting to a Changing World

Incidentally, energy *is* facing a shift no less radical than the invention of steam power. Just as steam power heralded the age of fossil fuels, we are now nearing the end of this era. The global stock in oil is slowly but steadily depleting (Heinberg, 2005). BP (2014) estimates the proved reserves of oil at around 1.7 trillion barrels. Proved reserves refer to "those quantities that [...] can be recovered in the future from

---

[1]`http://www.bmbf.de/en/19955.php`, accessed on April 13, 2015.

known reservoirs under existing economic and operating conditions."
(BP, 2014, p. 6). While technological progress and rising oil prices
may increase this amount in the future, these reserves would last for
about fifty years at the current consumption level of about 33 billion
barrels per year. However, global oil consumption is still growing.
Given the 2012/2013 growth rate of 1.4 percent (BP, 2014), proved
reserves may last for less than forty years. The prospects for natural
gas and coal are slightly less dire. Nevertheless, we are likely to see
a tremendous change in the accessibility and quantity of fossil fuels
within the first half of this century.

While fifty years may appear to be a long time, energy infrastructure
is built to last. Campbell (2012) notes that the average ages of U.S.
power plants and of components for transmission and distribution
exceed 30 and 40 years, respectively. Hence, a changing composition
of energy resources must be addressed well in advance.

The depletion of fossil fuels is not the only development affecting the
energy industry. The overwhelming scientific consensus on the human
contribution to climate change (Cook et al., 2013; Tol, 2014; Cook
et al., 2014) outlines the consequences of an excessive reliance on
carbon-based energy generation. In 2010, almost two thirds of global
CO_2-emissions were caused by two sectors—*electricity & heat* and
transportation (IEA, 2012). However, Davis et al. (2010) show that
the sources of the most threatening emissions over the next fifty years
are yet to be built. If these new generators and vehicles are powered
by carbon-neutral resources, the most drastic consequences of climate
change may still be averted.

For the past decades, nuclear fission has been considered one such
carbon-neutral power source. However, the Fukushima incident in
2011 has brought the associated risks back into the public mind and
eroded support for this form of energy generation in several countries
(Siegrist & Visschers, 2013; Hartmann et al., 2013). The storage of
nuclear waste that accumulates at the end of the fuel cycle is another
issue facing strong public opposition and fear (Slovic et al., 1991; Rosa

et al., 2010; Dawson & Darst, 2006). It should also be kept in mind that the global reserves of uranium and other nuclear fuels are, while quite vast, not infinite (OECD NEA & IAEA, 2014).

As a result, the focus is increasingly shifting towards renewable energy sources—water, wind, and sun. In 2013, hydroelectricity supplied 6.7 percent of global primary energy consumption, with other renewable sources (including biofuels) contributing another 2.2 percent (BP, 2014). Given the immense geological requirements that need to be met for the construction of hydroelectric dams, a transformation towards a renewable energy paradigm must be carried by wind and solar power.

1.3 The Transition towards Renewable Energy

The political will to realize such a change can be observed in several countries around the globe (REN21, 2014). Germany has heavily subsidized renewable generation through the *Erneuerbare-Energien-Gesetz* (Act on granting priority to renewable energy sources)[2] and decided to phase out nuclear power. The IEA (2013a) outlines further political measures to foster a more sustainable energy supply in several countries including the Netherlands, Norway and Sweden, Spain, and the United States. China has become the world leader in renewable energy generation (mainly from hydroelectric and wind power) and in supplying solar panels and wind turbines (Mathews & Tan, 2014). Yet, global subsidies to fossil fuels still exceeded those to renewable energy sources by a factor of six in 2011 (IEA, 2013b), a development largely driven by fixed gasoline prices in developing countries. This emphasizes the need for affordable clean energy technologies that combat poverty in those countries while not exacerbating the energy crisis (Detchon & van Leeuwen, 2014).

---

[2]https://www.clearingstelle-eeg.de/files/node/8/EEG_2012_Englische_Version.pdf, accessed on April 13, 2015.

However, large-scale integration of wind and solar power still faces immense challenges on a technical level. On the one hand, large power plants are replaced by a multitude of small generators, spread out over vast regions or entire countries. This decentralization of power generation places new strains on the power systems (dena GmbH, 2012; Paatero & Lund, 2007), which had originally been designed to transport energy from power plants through transmission and distribution grids to the consumers in a top-down fashion. On the other hand, solar and wind power crucially depend on exogenous factors, namely sun and wind. Consequently, supply cannot be adjusted to changes in demand as dynamically as it used to and the stability of energy grids is threatened. Hence, methods that enable a flexible management of energy demand are increasingly required (Gellings, 1985; Palensky & Dietrich, 2011; Strbac, 2008).

Addressing these issues is a *crucial requirement* for a sustainable energy supply. A promising approach is to embed information technology (IT) in the power system to enable intelligent control and management mechanisms—making energy *smarter*. Such *smart grids* (Farhangi, 2010; Amin & Wollenberg, 2005) are the main focus of this thesis and will be further introduced in the next section.

2 Smart Grids

The BDEW (German Association of Energy and Water Industries) defines a smart grid as "an energy network which integrates the consumption and feed-in behaviour of all market participants connected to it. It represents an economically efficient, sustainable power supply system with low losses and a high level of availability" (BDEW, 2013, p. 12). The integration is provided by information technology that automatically collects data on energy consumption and generation in regular intervals through *smart metering* devices. Coordination mechanisms subsequently use this data to determine the optimal configuration of the grid and effect adjustments in the power system accordingly. These effects can either be achieved through direct control mechanisms, i.e. by automatically (de)activating certain remote-controllable loads or generators, or indirectly through, for instance, price signals. Katz et al. (2011) further specify means to flexibly adjust demand through the concepts of *slack* and *slide*. They refer to slack as the potential of an energy load to change its consumption pattern without affecting operations or outcomes—thereby effectively storing energy. One example they outline concerns refrigerators that essentially store cooling energy. By dynamically adjusting the cooling cycles, refrigerators can react to variations in the energy supply while still keeping food sufficiently cooled. Slide on the other hand describes the explicit rescheduling of certain tasks, such as running a washing machine, to off-peak times.

2.1 The Smart Grid Architectural Model

Research on smart grids has received extensive political support and funding around the globe, which may be best exemplified by the ARPA-E agency within the U.S. Department of Energy[3] and the Smart Grid Task Force of the European Commission[4]. These institutions advise policymakers and fund projects on smart grids as part of a larger agenda towards a more sustainable energy paradigm. Within Germany, several lighthouse projects funded through the federal E-Energy initiative[5] have also provided insights on challenges and solutions in respect to smart grid implementations. To guide future smart grid research projects, the CEN-CENELEC-ETSISmart Grid Coordination Group (2012) have compiled the Smart Grids Architectural Model (SGAM), illustrated in Figure I–1, a reference model to categorize smart grid processes, products, and research.

---

[3]http://arpa-e.energy.gov/, accessed on April 13, 2015.
[4]http://ec.europa.eu/energy/en/topics/markets-and-consumers/ smart-grids-and-meters/smart-grids-task-force, accessed on April 13, 2015.
[5]http://www.e-energy.de/en/, accessed on April 13, 2015.

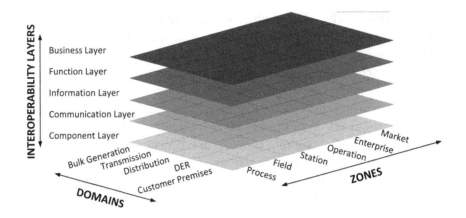

Figure I–1: Smart Grid Architectural Model (adapted from CEN-
CENELEC-ETSISmart Grid Coordination Group, 2012)

The foundation of the SGAM is the smart grid plane, which represents electrical processes and the associated management procedures within a smart grid. The electrical energy conversion chain is captured by its various **domains**. Traditionally, electrical energy has been produced in large power plants (*bulk generation*) and was successively transmitted through *transmission* and *distribution* grids to the *customer premises*. The rise of wind and solar power has changed this straightforward chain through the introduction of small decentralized generators. These distributed energy resources (*DER*) feed energy into the grid at the distribution level.

The **zone** dimension describes the hierarchical levels of power systems management within a smart grid, based on the concepts of aggregation and functional separation (CEN-CENELEC-ETSISmart Grid Coordination Group, 2012). Aggregation refers to both the data and spatial perspectives. Equipment involved in the electrical *processes* is monitored at the *field* level through intelligent electronic devices. The *station* zone aggregates data from spatially distributed field equipment to reduce the overall amount of data that needs to be processed at the operational level. The *operation* zone controls the power system in the respective domain through, for instance, energy management systems. The *enterprise* level "includes commercial and organizational processes, services and infrastructures" (CEN-CENELEC-ETSISmart Grid Coordination Group, 2012, p. 29) of businesses involved in the energy sector, such as utility companies or ancillary service providers. Finally, the *market* zone contains market operations in each domain. The level of aggregation necessary in each zone follows directly from the function of the zone. For instance, data at the field level needs to be available in detail and at a high temporal resolution to secure the stability and functionality of the power equipment in the respective domain. The enterprise level, on the other hand, generally uses data that was spatially and temporally aggregated to support business processes and decisions.

The smart grid plane is covered by five **interoperability layers**, which ensure interoperability across all systems and components within a smart grid. The *component layer* includes actors, applications, as well as IT and power systems equipment. The *communication layer* contains protocols and mechanisms that enable an exchange of information. A major focus of research in this area concerns the further development of specification according to existing communication standards (CEN-CENELEC-ETSISmart Grid Coordination Group, 2012). The *information layer* describes which information is required or provided by specific smart grid elements. The service or process that is enabled by a certain smart grid concept—independent from its physical implementation—is captured in the *function layer*. The *business layer* contains business objectives and processes, including issues concerning economic and regulatory policy.

The SGAM constitutes a comprehensive architectural model for smart grid innovation and research. In Section 3, it will serve as a framework to allocate the papers included in this thesis within the context of broader research on smart grids.

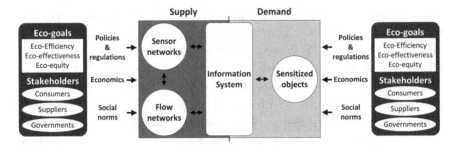

Figure I–2: Energy Informatics Framework (as in Watson et al., 2010)

2.2 Energy Informatics

Early smart grid research has been largely driven by the electrical engineering and computer science disciplines, developing the technical foundations of smart grid technologies. However, with an increasing political interest in implementing smart grids, research efforts on the behavioral, social, and economic consequences of an increasing digitalization of the power grid have also intensified. The intersection of technological progress and these fields of study is the research locus of the Information Systems (IS) discipline. In 2010, Watson et al. published a call for action to intensify IS research on topics surrounding smart grids, renewable energy, and climate change. They outlined an agenda for *Energy Informatics* research, which seeks to employ, adapt, and further develop IS methods and insights to contribute to transdisciplinary efforts towards a sustainable energy paradigm. Through this call and the *Energy Informatics Framework* (EIF) they proposed, Watson et al. (2010) have provided an anchor for smart grid research within the IS community.

As outlined in Figure I–2, the EIF represents a high-level framework for research projects within the Energy Informatics domain. The essential components of the framework are *flow networks*, such as transmission grids, streets, and pipelines; a *sensor network* that provides data on physical items or environmental conditions to determine the optimal configuration of a flow network; and *sensitized objects*, which can sense and report data about their use and may be remote-controllable. The *information system* at the center of the framework builds upon these components to align supply and demand in the power system subject to certain *eco-goals*: efficiency, effectiveness, and equity. These goals are set by the *stakeholders* of the system, who also influence it through policies, social norms, and economic considerations.

Several other frameworks and research models concerning Green IS and Energy Informatics have been introduced to stimulate and guide research, both before and after the EIF. For instance, the "Belief-Action-Outcome Framework for IS and Sustainability" by Melville (2010) captures how individual beliefs about the environment, as well as societal and organizational structure, influence individual actions, which in turn affect the behavior of organizations and society. The frameworks by Elliot (2011), as well as by Hovorka & Corbett (2012), both emphasize the inherent interdisciplinarity of research on sustainability.

This theoretical groundwork has resulted in a heightened interest in sustainability and energy challenges within the IS community. However, research is often focused on the outer layers of the EIF in Figure I–2, i.e. eco-goals and stakeholders), thus frequently investigating the organizational effects of *green* information systems. For instance, Corbett (2013) analyzes the design and use of carbon management systems to promote ecologically responsible behavior within organizations. Seidel et al. (2014) outline the role of IT in enabling organizational transformations towards sustainability. Loock et al. (2013) present a study on how feedback affects energy-efficient behavior. The next section outlines how this thesis complements and contributes to this growing body of Energy Informatics literature.

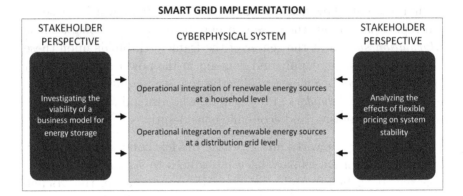

Figure I–3: Contribution of thesis to a smart grid implementation

3 Research Framework

This thesis builds upon the Energy Informatics Framework to investigate research aspects necessary for a successful smart grid implementation. Figure I–3 illustrates how these aspects reflect concepts within the EIF and how they contribute to a comprehensive smart grid solution. This stylized version of the EIF (c.f. Figure I–2) outlines on the one hand that the thesis addresses the cyberphysical system at the center of the Energy Informatics Framework, which links the information system to energy networks, sensor network, and sensitized objects. Through the design and evaluation of two IT artifacts, the operation of this cyberphysical system is improved and the integration of renewable energy sources is enabled. While one artifact considers an individual household that owns a roof-top photovoltaic installation as well as an electric vehicle, the other controls the battery charging processes for a microgrid located at a U.S. army base with vast photovoltaic generation.

On the other hand, this analysis of the cyberphysical system is complemented by the perspectives of the stakeholders invested in the power system. A business model for the aggregation of electric vehicles as energy storage is evaluated critically. Thereby, insights on the business aspect of smart grid technologies are provided to managers and policymakers. Another aspect that is investigated in this thesis concerns system instability as a result of price signals in flexible energy tariffs. Possible adverse effects arising from the increasing automation within a smart grid are theoretically modeled and the implications for demand side management mechanisms are outlined.

Each of these four aspects is presented as a building block that describes an independent contribution to an improved energy system. Together, they provide a nuanced assessment of possible routes towards a large-scale smart grid implementation. This common goal is emphasized by three broad research themes that guide each approach.

Scale. Power grids cover entire nations and continents while simultaneously branching into hundreds of thousands of power lines and circuits, down to every last electric appliance. Relating to the domain and zone dimensions in the SGAM (c.f. Figure I–1), the crucial question is at what scale or at what location within the smart grid plane *smartness* pays off. Individual devices and buildings often have very erratic consumption patterns, which are difficult to predict and optimize. Nevertheless, local solutions may outperform large-scale approaches due to reduced complexity or efficiency losses from aggregation.

Technology. Following the zone dimension of the SGAM, specific information management mechanisms—IT artifacts, business models, and market designs—are investigated. Specific showcases on how information technology improves the power grid are presented. Thereby, design theory (e.g. Hevner et al., 2004; Jenkin et al., 2011) is applied and extended.

Economics. Hahn & Stavins (1992) outline the role of economic incentives to foster environmental protection. The main case for smart grids is environmental as well—increased energy efficiency, the integration of renewable energy sources, reduced reliance on fossil fuels. Hence, the papers included in this thesis seek to provide insights on the business layer of the SGAM. Throughout the following chapters, concepts are evaluated using real-world data, if possible, to estimate their economic and ecological benefits. After all, even the best idea will not evolve beyond an idea if these benefits cannot be credibly illustrated.

Each of the four building blocks described above is covered in a research paper that enters as a separate chapter into this thesis. While the papers approach the research themes—scale, technology, and economics—in different ways, the Smart Grid Architectural Model provides a common framework for all of them. Table I–1 illustrates how the chapters match the dimensions of the SGAM, outlining a cohesive structure. The **technological concepts** in each chapter relate to the function layer because a specific application is introduced that delivers a particular service. Since Chapters II and III include the design of IT-artifacts, the information and component layers are also relevant. While Chapter IV focuses on the development of a business model, the associated revenue streams are simulated and require a thorough understanding of the information and component layers, as well. Chapter V analyzes market design issues for demand side management on a theoretical level, thus focusing on the function layer.

Furthermore, each chapter relates to the business layer because the issue of **economic incentives** of smart grid technologies is investigated. In Chapters II, III, and IV revenues that can be attributed to the IT-artifact or business model are compared to the associated cost structures. Chapter V, on the other hand, analyzes the effect of price signals.

While the issues of technology and economics unify the papers included in this thesis, the issue of **scale** provides them with structure. Table I–1 shows that each chapter considers a different combination of SGAM zones and domains. These combinations reflect different levels of scale, as visualized in Figure I–4. As Chapter II develops an operational artifact for individual households, domains are limited to the customer

Table I–1: Allocation of research papers within SGAM

		Chapter 2	Chapter 3	Chapter 4	Chapter 5
Layers	Business Layer	•	•	•	•
	Function Layer	•	•	•	•
	Information Layer	•	•	•	
	Communication Layer				
	Component Layer	•	•	•	
Domains	Bulk Generation				
	Transmission			•	•
	Distribution		•		•
	DER	•	•		•
	Customer Premises	•	•	•	•
Zones	Market				•
	Enterprise			•	
	Operation	•	•	•	
	Station				
	Field				
	Process				•

(household) premises and DER through the photovoltaic installation. Chapter III scales a similar concept up to the microgrid level, which is large enough to have an impact on the distribution grid. The business model introduced in Chapter IV trades energy in the market for frequency regulation, which is organized at the transmission level. This market perspective at a macro level is further investigated in Chapter V for demand side management mechanisms.

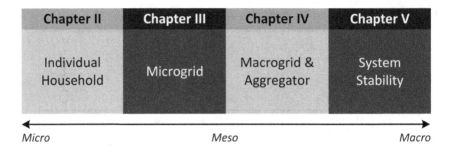

Chapter II	Chapter III	Chapter IV	Chapter V
Individual Household	Microgrid	Macrogrid & Aggregator	System Stability

Micro   *Meso*   *Macro*

Figure I–4: Structure of thesis according to scale of applications

The research papers included in this thesis each analyze a specific application of information technology in a smart grid. These applications are evaluated with respect to economic incentives and provide insights on the potential of smart grid technologies at varying levels of scale. Each chapter is briefly summarized below.

4 Thesis Summary

4.1 Summary: Designing IT Solutions for Individual Households

Chapter II introduces a showcase to demonstrate how the process of designing IT applications for smart grids differs from traditional IS design. It extends design theory by emphasizing the cyberphysical nature of smart grids and possible interferences from the legacy power grid that could arise during the design process. The showcase, which was presented at the *2013 International Conference on Information Systems* and received the Best Conference Theme Paper Award, focuses on electric vehicles (EVs) as energy storage devices within a smart grid. These, together with the IT artifact, constitute the component layer. On average, cars are used only four percent of the day for driving and are parked during the remainder. EVs could be used during these idle times to compensate the fluctuations in energy generation from renewable sources. The vehicle would charge its battery during times of high generation and discharge it during energy scarcity using vehicle-to-grid technology. The revenues from this service contribute to reducing the price difference between electric and conventional vehicles.

The chapter introduces an IT artifact for individual households that own an EV and a photovoltaic panel—hence, the focus on the customer premises domain. At the functional layer, the artifact intelligently charges and discharges the vehicle depending on photovoltaic generation and the energy demand of the household. Since predicting energy demand and mobility behavior on such a micro-level for optimization is very difficult, the artifact implements a decision strategy, which uses a flexible threshold that reflects information on historical driving patterns.

To asses the economic feasibility, the artifact is evaluated using data on photovoltaic generation from Bavaria and mobility patterns from a large study in Germany. For the sample household, the artifact reduces annual energy costs by 258 EUR, which equals about two to three monthly electricity bills for the average German 4-person household. Over the lifetime of an electric vehicle, it also far exceeds the effect of federal EV subsidies in the U.S. Additionally, the artifact allows more photovoltaic energy to be used locally by the household, decreasing the strain on the grid from feed-ins.

The chapter closes by outlining smart grid design guidelines that complement those proposed by Hevner et al. (2004) for general IS design. These concern understanding the legacy system, the impact of network effects, as well as the role of the end-user.

4.2 Summary: Designing an Energy Information System for Microgrid Operation

Chapter III investigates how the volatility of renewable energy sources can be managed at the level of a microgrid. Microgrids are semi-autonomous energy grids that are connected to the macrogrid through a coupling point and may contain distributed energy generation, storage, and demand. For instance, the power system in residential areas consisting of multiple households or in large office buildings may be designed as a microgrid. In this study, an IT artifact for microgrid operation at a U.S. army base is developed. The study was completed during a research visit at Lawrence Berkeley National Laboratory (LBNL) in Berkeley, California, and presented at the *2014 International Conference on Information Systems*.

In this context, the customer premise is the army base, a training facility whose power demand fluctuates between one and two megawatts for most of the time. Initially, the base had contained a two megawatt-peak photovoltaic installation, which was expanded by an additional megawatt during the study and is projected to eventually reach eight megawatt-peak. However, the agreement with the utility company limits the power export to one megawatt. Hence, a one megawatt-hour battery was installed at the base to store excess energy. Since the PV installation can only be turned off in discrete segments of half a megawatt if energy exports exceed the limit, the IT artifact is needed to intelligently (dis)charge the battery to limit these incidents.

The function of the IT artifact is to optimize the charging schedule of the battery using the software DER-CAM, which has previously been developed at LBNL. In this chapter, forecasters for the energy demand (load) of the army base and the generation of photovoltaic energy are developed. These forecasters provide DER-CAM with the input necessary to determine the optimal schedule. Since the load curve of the base exhibits strong repetitive patterns, the load forecaster employs a Discrete Fourier Transformation to estimate future demand. The forecaster for photovoltaic generation uses data on the altitude of the sun and the estimated cloud cover as provided by a weather service.

The prototype of the artifact is evaluated using historical data on energy demand, photovoltaic generation, and weather conditions provided by the base. The performance of the IT artifact is compared to a decision strategy similar to the one used in Chapter II and to a situation without battery. The optimization conducted by the artifact outperforms both alternatives since the strong demand patterns of the base enable good forecasts. However, from an economic perspective, battery prices are still too high for the reduced energy costs to compensate initial investments. Nevertheless, the ecological benefit is noteworthy since the artifact enables the generation of additional six megawatt-hours of clean energy during the test week, which is close to the theoretical maximum.

4.3 Summary: A Business Model for Employing Electric Vehicles for Energy Storage

Chapter IV scales the concept of using electric vehicles for energy storage (EV4ES) up to a macrogrid level. It introduces an aggregator that sells energy at the market for frequency regulation, which is located at the transmission level of the grid. It extends related work in this area by developing and evaluating a comprehensive business model. The chapter is based on research presented at the *2012 International Conference on Information Systems* and at the *2013 European Conference on Information Systems.*

The business case is developed for a company that operates multiple parking facilities. There are several advantages of considering such a company as an EV4ES aggregator. It already has access to a vast amount of cars that park in the facilities throughout the day and can rely on an existing billing infrastructure. Furthermore, if the number of electric vehicles continues to increase, parking garage operators may need to install charge points at their facilities irrespective of whether or not they offer an EV4ES service. Nevertheless, the core question of this chapter is if and under what conditions such a business model is profitable.

The main revenue stream (selling energy at the market for frequency regulation) and its functional implementation within the garages are simulated using extensive real-world data sets. Energy is sold at the German market for frequency regulation based on auction data from 2014. The daily fluctuations in the occupancy of the parking facilities are modeled after data gathered from several German parking garages. Furthermore, the probability distribution for the state of charge of the battery when entering the facility is based on data from EVs in the Netherlands. The simulation calculates the annual revenue from energy sales for a given number of vehicles entering the garages on a daily basis.

A realistic estimate of the resulting revenues ranges between 150,000 and 200,000 EUR in 2014, given that 10,000 vehicles enter the parking garages per day. These revenues fail to compensate the substantial investments into charge points and IT infrastructure necessary to implement an EV4ES service. Even if investments were covered by subsidies or a broader business strategy, simply selling energy to EV owners results in substantially higher revenues than ancillary services such as frequency regulation. Hence, the chapter concludes by pointing out that the challenge, which a widespread adoption of electric mobility poses to the grid, may be fundamentally underestimated since economic factors are not sufficiently taken into account.

4.4 Summary: Mechanism Stability—Flash Crashes and Avalanche Effects

The final chapter of this thesis discusses how an increasing automation of coordination mechanisms within a system may lead to instabilities. This problem is outlined through two concepts that reflect a sudden imbalance of demand and supply—*flash crashes* in financial markets and *avalanche effects* in demand side management (DSM) mechanisms. Since flash crashes can already be observed in practice, the theoretical development within the chapter is based on a financial market. The implications for DSM are outlined at the end of the chapter. The work in this chapter is based on a paper presented at the *2014 Hawaii International Conference on System Sciences*, which was nominated for a Best Paper Award, and a subsequent paper published in the *Journal of Management Information Systems*.

Flash crashes are characterized by sudden price drops that are almost immediately recovered. Chapter V explains the occurrence of flash crashes as the result of technological progress using Hypergame Theory. Essentially, it is argued that games sometimes turn from sequential to simultaneous while traders continue to pursue the strategies from a sequential game. This results, for instance, in an unwarranted amount of simultaneous sales which can be further amplified by herd behavior. Once traders realize their mistake, the price returns to the appropriate level, resulting in the characteristic flash crash spike.

DSM mechanisms face a similar phenomenon in avalanche effects. If smart devices, such as laundry machines or dryers, are triggered by a price signal, an oversupply of power can suddenly turn into a lack of power at the household, distribution, or transmission levels if too many devices react simultaneously. Chapter V concludes by noting that the problem could be alleviated by updating price signals very frequently. However, as this may induce an increase in communication costs for the utility company, the feasibility of DSM mechanisms is uncertain.

5 Conclusion

The world economy is fundamentally dependent on fossil fuels. The transition to a sustainable energy paradigm based on renewable resources is a challenge for a generation. Smart grids *will not be* the universal solution to this grand challenge. However, the results of the research included in this thesis show that smart grids *can contribute to this solution*. Nevertheless, it also requires a collective will to innovate and to invest, to forfeit a part of the prosperity today so that future generation can prosper. To conclude, the key insights are summarized below and an outlook on possible future research avenues is given.

5.1 Key Insights

The key insights are outlined according to the major research themes of this thesis—scale, technology, and economics. Additionally, policy implications are discussed.

Scale

The research presented in this thesis demonstrates that information technology can improve efficiency and operation of the power grid if coordinated at the micro and meso levels. However, these IT artifacts or systems need to reflect the features specific to each level. For instance, individual households exhibit very volatile patterns of energy consumption—turning on the oven or microwave may result in tremendous spikes in demand for short periods of time. The same holds true for mobility behavior, since daily and weekly routines are not set in stone. Predicting this behavior is almost impossible and results in a high degree of uncertainty. In Chapter II, this issue is addressed by using a plug-in hybrid EV for energy storage, such that the owner does not need to delay trips even when the battery is unexpectedly empty. Furthermore, the electric vehicle as a local storage device is matched with a local intermittent generator. Exploiting the synergies between local generation and storage will certainly contribute to a better integration of distributed energy resources without relying on large-scale aggregation schemes.

The microgrid combines smoother and more predictable consumption patterns with the advantages of local generation and storage. By optimizing loads and storage within the microgrid, subject to constraints on energy exports to the macrogrid, it may be the most promising way to integrate renewable energy sources into the existing power system. A major obstacle to this goal is the coordination and distribution of investments and revenues among different stakeholders of the microgrid. While Chapter III investigates a microgrid with a single stakeholder, microgrids with multiple stakeholders (e.g. in residential areas) are more difficult to realize.

Technology

Chapter II shows that information technology for a smarter grid does not need to be complex or expensive to be effective. The IT artifact executes a very simple trigger strategy with limited hardware and software requirements. The substantial reduction in feed-ins also illustrates the benefits that can arise from synergies between various *green technologies* on a local level. However, technological requirements increase for the efficient management of a microgrid. Since the artifact in Chapter III optimizes the battery schedule, precise forecasts of energy demand and generation are needed. These require a regular supply of internal (load, generation) and external (weather) data.

For the macrogrid, as outlined in Chapter IV, necessary investments currently far outweigh possible benefits. These include charge points and ICT infrastructure for scheduling and billing systems to aggregate the storage capacities of multiple electric vehicles. Furthermore, Chapter V demonstrates that simple mechanisms become more problematic when multiple devices simultaneously react to the same signal.

Economics

By exploiting synergies between local storage and generation technologies, the IT artifacts introduced in this thesis provide a tangible economic benefit. The artifact presented in Chapter II provides a much stronger incentive to acquire an electric vehicle than the U.S. federal subsidy designed to achieve the same. Similarly, the energy costs for the microgrid in Chapter III are substantially reduced. However, neither artifact fully compensates the acquisition costs of an electric over a conventional vehicle in the first case and over a large stationary battery in the latter case. This emphasizes that improvements in energy storage technologies and cost reductions are essential to a successful integration of renewable energy sources.

Whether aggregators of electric vehicles will eventually contribute to this integration is highly uncertain, given the results of Chapter IV. The revenues from energy sales at the regulation market pale in comparison to the necessary investments. Furthermore, markets are currently not designed to allow providers with stochastic resources (such as EVs) to enter the market.

Policy

Addressing the lack of a market for stochastic energy storage is also one of the major policy implications from this thesis. While electric vehicles are generally able to lessen the volatility of solar and wind power, their primary purpose will continue to be mobility. Hence, even if aggregated, there will remain a certain small probability that the aggregator cannot supply the stipulated amount of energy because an unexpectedly large number of vehicles is driving instead of parking. If EVs are intended to provide large-scale storage for intermittent renewable power, markets need to be designed in a way that reflects this characteristic. Chapter IV also outlines that prices at these markets would need to be substantially above current levels.

Furthermore, Chapter II shows that policies need to better align the goals of increasing renewable generation and supporting energy storage solutions. The subsidy scheme for photovoltaics in place in Germany during the time of the study distorted prices in such a way to actively disincentivize storage solutions. Future policies need to reflect the inherent synergies and dependencies between renewable generation and energy storage.

5.2 Outlook

While the studies included in this thesis only focus on a few facets of smart grids and energy research, they provide a basis for several future research paths. On the one hand, this includes aspects that have been excluded in the concepts that were introduced; on the other hand, these concepts can be expanded and adapted to other settings.

The summaries in Section 4 show that further investigation is necessary with respect to the communication layer of the SGAM (cf. Figure I–1). The studies generally assume that smart metering infrastructure is already in place, as well as protocols that enable the transmission of the collected data. The interaction between the IT artifacts and the communication infrastructure needs to be further analyzed with respect to possible requirements one puts on the other.

While security and data privacy issues have not been the focus of this thesis, they constitute a very active research area. Cyber security is considered to be paramount to the realization of smart grids (Ericsson, 2010). The increasing interconnection of millions of electrical devices through information and communication technology opens up new vulnerabilities of the power system to outside attacks (Wang & Lu, 2013; Mo et al., 2012). Data privacy is a related aspect that emphasizes the personal nature of the data collected by smart meters and requires a balanced approach. The localized solutions presented in Chapters II and III may contribute to a solution of these issues and merit further investigation.

Chapter III also provides a starting point for expanding and adapting the concepts introduced in tis thesis. As the microgrid that was discussed belongs to a single stakeholder, future studies should investigate the effect of introducing multiple stakeholders. For instance, these could be the homeowners living in a residential area. Since they are likely to be heterogeneous with respect to their energy consumption and their abilities to generate and store energy, a mechanism that allocates the benefits of the microgrid needs to be devised. Another issue that merits discussion is the market inclusion of multiple mcirogrids into the macrogrid. The showcase presented in Chapter III assumes an export limit imposed by the utility company. An increase in the number of microgrids might open up further options, such as time-varying limits or the microgrid entering wholesale energy markets.

Chapter IV puts the current economic viability of aggregating electric vehicles for energy storage in question. Nevertheless, EVs continue to be a promising way of balancing the intermittency of renewable energy sources. Future research should focus on which market conditions and designs support business models that seek to exploit the energy storage provided by EVs and how economic feasibility can be achieved.

Finally, the incentives to adopt smart grid technologies—or any sustainable energy technology—are determined by the policies that shape the economic environment. It has been previously noted that subsidies to encourage renewable energy investments should not do so at the expense of energy storage. Both technologies are necessary for a successful transition to a sustainable energy paradigm and policies need to be designed in a way that reflects this. Furthermore, even though the IT artifacts designed in this thesis provide substantial financial benefits, they cannot fully compensate the necessary investments into electric vehicles and standalone batteries, respectively. Hence, policy research will particularly need to analyze how the development and dissemination of energy storage solutions can be supported.

6 References

Amin, S. M. & Wollenberg, B. F. (2005). Toward a smart grid: power delivery for the 21st century. *IEEE Power and Energy Magazine*, *3*(5), 34–41.

Armaroli, N. & Balzani, V. (2007). The future of energy supply: Challenges and opportunities. *Angewandte Chemie (International ed. in English)*, *46*(1-2), 52–66.

BDEW. (2013). *BDEW Roadmap: Realistic Steps for the Implementation of Smart Grids in Germany*. Bundesverband der Energie und Wasserwirtschaft.

BP. (2014). *BP Statistical Review of World Energy 2014*. London: BP p.l.c.

Burton, F. D. (2009). *Fire: The Spark That Ignited Human Evolution*. Albuquerque: University of New Mexico Press.

Campbell, R. J. (2012). *Weather-Related Power Outages and Electric System Resiliency*. CRS Report for Congress.

CEN-CENELEC-ETSISmart Grid Coordination Group. (2012). *Smart Grid Reference Architecture*. Retrieved on April 13, 2015, from `http://ec.europa.eu/energy/sites/ener/files/documents/xpert_group1_reference_architecture.pdf`

Chandler, A. D. (1992). Organizational Capabilities and the Economic History of the Industrial Enterprise. *The Journal of Economic Perspectives*, *6*(3), 79–100.

Cook, J., Nuccitelli, D., Green, S. A., Richardson, M., Winkler, B., Painting, R., ... Skuce, A. (2013). Quantifying the consensus on anthropogenic global warming in the scientific literature. *Environmental Research Letters*, *8*(2), 024024.

Cook, J., Nuccitelli, D., Skuce, A., Jacobs, P., Painting, R., Honeycutt, R., ... Way, R. G. (2014). Reply to 'Quantifying the consensus on anthropogenic global warming in the scientific literature: A re-analysis'. *Energy Policy, 73*, 706–708.

Corbett, J. (2013). Designing and Using Carbon Management Systems to Promote Ecologically Responsible Behaviors. *Journal of the Association for Information Systems, 14*(7), Article 2.

Davis, S. J., Caldeira, K. & Matthews, H. D. (2010). Future CO2 emissions and climate change from existing energy infrastructure. *Science, 329*(5997), 1330–1333.

Dawson, J. I. & Darst, R. G. (2006). Meeting the challenge of permanent nuclear waste disposal in an expanding Europe: Transparency, trust and democracy. *Environmental Politics, 15*(4), 610–627.

dena GmbH. (2012). *dena Distribution Grid Study (in German)*. Berlin: Deutsche Energie-Agentur GmbH. Retrieved on April 13, 2015, from http://www.dena.de/en/projects/energy-systems/dena-distribution-grid-study.html

Detchon, R. & van Leeuwen, R. (2014). Policy: Bring sustainable energy to the developing world. *Nature, 508*(7496), 309–311.

Doménech, J. L., Gil-Pérez, D., Gras-Martí, A., Guisasola, J., Martínez-Torregrosa, J., Salinas, J., ... Vilches, A. (2007). Teaching of Energy Issues: A Debate Proposal for a Global Reorientation. *Science & Education, 16*(1), 43–64.

Elliot, S. (2011). Transdisciplinary Perspectives on Environmental Sustainability: A Resource Base and Framework for IT-Enabled Business Transformation. *Management Information Systems Quarterly, 35*(1), 197–236.

Ericsson, G. N. (2010). Cyber Security and Power System Communication—Essential Parts of a Smart Grid Infrastructure. *IEEE Transactions on Power Delivery, 25*(3), 1501–1507.

Farhangi, H. (2010). The path of the smart grid. *IEEE Power and Energy Magazine, 8*(1), 18–28.

Gellings, C. W. (1985). The concept of demand-side management for electric utilities. *Proceedings of the IEEE*, *73*(10), 1468–1470.

Hahn, R. W. & Stavins, R. N. (1992). Economic Incentives for Environmental Protection: Integrating Theory and Practice. *The American Economic Review*, *82*(2), 464–468.

Hartmann, P., Apaolaza, V., D'Souza, C., Echebarria, C. & Barrutia, J. M. (2013). Nuclear power threats, public opposition and green electricity adoption: Effects of threat belief appraisal and fear arousal. *Energy Policy*, *62*, 1366–1376.

Heinberg, R. (2005). *The Party's Over: Oil, War and the Fate of Industrial Societies* (2nd ed.). Gabriola Island, BC: New Society Publishers.

Hevner, A., March, S., Park, J. & Ram, S. (2004). Design Science in Information Systems Research. *Management Information Systems Quarterly*, *28*(1), 75–105.

Höller, J., Tsiatsis, V., Mulligan, C., Avesand, S., Karnouskos, S. & Boyle, D. (2014). *From Machine-to-Machine to the Internet of Things*. Amsterdam and Boston: Academic Press.

Hovorka, D. & Corbett, J. (2012). IS Sustainability Research: A trans-disciplinary framework for a 'grand challenge'. *ICIS 2012 Proceedings*, Paper 8.

IEA. (2012). *CO2 Emissions from Fuel Comubstion*. Paris: International Energy Agency.

IEA. (2013a). *Energy Policy Highlights*. Paris: International Energy Agency.

IEA. (2013b). *Redrawing the Energy-Climate Map: Executive Summary*. Paris: International Energy Agency. Retrieved on April 13, 2015, from `http://www.iea.org/media/freepublications/weo/WEO2013_Climate_Excerpt_ES_WEB.pdf`

Jenkin, T. A., Webster, J. & McShane, L. (2011). An agenda for 'Green' information technology and systems research. *Information and Organization*, *21*(1), 17–40.

Katz, R. H., Culler, D. E., Sanders, S., Alspaugh, S., Chen, Y., Dawson-Haggerty, S., ... Shankar, S. (2011). An information-centric energy infrastructure: The Berkeley view. *Sustainable Computing: Informatics and Systems*, *1*(1), 7–22.

Le Billon, P. & El Khatib, F. (2004). From Free Oil to 'Freedom Oil': Terrorism, War and US Geopolitics in the Persian Gulf. *Geopolitics*, *9*(1), 109–137.

Loock, C.-M., Staake, T. & Thiesse, F. (2013). Motivating Energy-Efficient Behavior with Green IS: An Investigation of Goal Setting and the Role of Defaults. *Management Information Systems Quarterly*, *37*(4), 1313–1332.

Mathews, J. A. & Tan, H. (2014). Economics: Manufacture renewables to build energy security. *Nature*, *513*(7517), 166–168.

Melville, N. (2010). Information Systems Innovation for Environmental Sustainability. *Management Information Systems Quarterly*, *34*(1), 1–21.

Mo, Y., Kim, T. H.-J., Brancik, K., Dickinson, D., Lee, H., Perrig, A. & Sinopoli, B. (2012). Cyber–Physical Security of a Smart Grid Infrastructure. *Proceedings of the IEEE*, *100*(1), 195–209.

OECD NEA & IAEA. (2014). *Uranium 2014: Resources, Production and Demand*. OECD Nuclear Energy Agency.

Paatero, J. V. & Lund, P. D. (2007). Impacts of Energy Storage in Distribution Grids with High Penetration of Photovoltaic Power. *International Journal of Distributed Energy Resources*, *3*(1), 31–45.

Palensky, P. & Dietrich, D. (2011). Demand Side Management: Demand Response, Intelligent Energy Systems, and Smart Loads. *IEEE Transactions on Industrial Informatics*, *7*(3), 381–388.

Pimentel, D. & Pimentel, M. H. (2008). *Food, Energy, and Society* (3rd ed.). Boca Raton: CRC Press.

REN21. (2014). *Renewables 2014 Global Status Report*. Paris: REN21 Secretariat.

Rosa, E. A., Tuler, S. P., Fischhoff, B., Webler, T., Friedman, S. M., Sclove, R. E., ... Short, J. F. (2010). Nuclear waste: knowledge waste? *Science*, *329*(5993), 762–763.

Ross, M. (2006). A Closer Look at Oil, Diamonds, and Civil War. *Annual Review of Political Science*, *9*(1), 265–300.

Seidel, S., Recker, J., Pimmer, C. & Vom Brocke, J. (2014). IT-enabled Sustainability Transformation—the Case of SAP. *Communications of the Association for Information Systems*, *35*, Article 1.

Siegrist, M. & Visschers, V. H. (2013). Acceptance of nuclear power: The Fukushima effect. *Energy Policy*, *59*, 112–119.

Slovic, P., Flynn, J. H. & Layman, M. (1991). Perceived risk, trust, and the politics of nuclear waste. *Science*, *254*(5038), 1603–1607.

Strbac, G. (2008). Demand side management: Benefits and challenges. *Energy Policy*, *36*(12), 4419–4426.

Tarciso Borges, A. & Gilbert, J. K. (1999). Mental models of electricity. *International Journal of Science Education*, *21*(1), 95–117.

Tol, R. S. (2014). Quantifying the consensus on anthropogenic global warming in the literature: A re-analysis. *Energy Policy*, *73*, 701–705.

Wang, W. & Lu, Z. (2013). Cyber security in the Smart Grid: Survey and challenges. *Computer Networks*, *57*(5), 1344–1371.

Watson, R., Boudreau, M.-C. & Chen, A. (2010). Information Systems and Environmentally Sustainable Development: Energy Informatics and New Directions for the IS Community.

White, L. A. (1943). Energy and the Evolution of Culture. *American Anthropologist*, *45*(3), 335–356.

Wrangham, R. W. (2010). *Catching fire: How cooking made us human.* London: Profile Books.

II Designing IT Solutions for Individual Households

Working Paper. Parts of the paper have been published in:

Tobias Brandt, Stefan Feuerriegel, and Dirk Neumann, "Shaping a Sustainable Society: How Information Systems Utilize Hidden Synergies between Green Technologies" (2013), ICIS 2013 Proceedings, Paper 7. Recipient of Best Conference Theme Paper Award.

Abstract

In recent years, research on the role of information systems (IS) in creating sustainable power systems has become a cornerstone of the IS discipline. In this paper we consider a particular aspect of this role—information systems as the facilitator of synergies between green technologies in the Smart Grid. Following a design perspective we propose that information systems can enhance technical objects with new features, thereby increasing their potential value towards a sustainable energy paradigm. We substantiate this proposition using a case study on batteries of electric vehicles. The information system designed in this study enables the battery to serve as an energy storage device in conjunction with a residential photovoltaic system while the vehicle is parked at home. This IS-enabled additional feature provides several financial and environmental benefits to the owner and society at large. Additionally, based on this case study, we outline possible interferences and obstacles in designing information systems for the Smart Grid and provide recommendations to guide future research.

1 Introduction

The pursuit of environmentally sustainable practices in society and economy has become one of the central objectives for many governments, companies, and citizens around the globe. Its incarnation takes various shapes—from organic farming in California to the phaseout of nuclear power plants in Germany and solar-powered rural electrification in Bangladesh. Through *Green Information Systems* and *Energy Informatics*, it has also developed into a major stream of Information Systems (IS) research. The scale and scope of publications range from theoretical groundwork (Elliot, 2011; Melville, 2010; Watson et al., 2010) through transitions towards sustainable information systems (Hilpert et al., 2013; Seidel et al., 2014) to design and behavior (Corbett, 2013; Loock et al., 2013). However, since "green" technologies are still facing tremendous obstacles with respect to costs and system integration, further research is clearly needed in order to overcome these barriers.

In this paper we investigate the impact of information systems at the intersection with physical power systems from a theoretical as well as an applied design perspective. This intersection is commonly referred to as the *Smart Grid* (Farhangi, 2010) and issues surrounding smart grids have been a cornerstone of recent Energy Informatics research (Goebel et al., 2014). We start out by developing the guiding hypothesis of our research—*information systems give access to new features of technical objects within the power system.* For this purpose, we introduce a theoretical model that relates design features of the information system to features of the power system. We coin them *Enhancing Features* since they enhance a technical object by providing a new use that was previously inaccessible. We proceed by presenting a case study on electric vehicles and residential renewable energy sources to substantiate the theoretical model. Based on this case study, the theoretical model, and on our experience in designing the information system, we outline possible obstacles and opportunities when designing information systems for the Smart Grid. We condense these insights into a set of guidelines to advise future IS design research with a smart grid background.

The remainder of this paper is structured as follows. In Section 2, we discuss research related to our work. We introduce our research model and develop the guiding hypothesis in Section 3. Section 4 contains the case study. We discuss implications from the case study relating to the achievement of anticipated outcomes through IS design in Section 5. Section 6 concludes with a summary of our analysis and the derived guidelines.

2 Related Work

In this section we provide an overview of publications related to our research. We divide this review into work related to the research domain (Energy Informatics) and to the research approach (Design Science). Since there exists a vast body of literature for each area, we focus on work closely relevant to our research. We conclude this section by identifying the research gap subsequently addressed in this paper.

2.1 Research Domain: Energy Informatics

Watson et al. (2010) define Energy Informatics (EI) as an interdisciplinary subfield of IS research "concerned with analyzing, designing, and implementing systems to increase the efficiency of energy demand and supply systems" (p. 24). The need for research on a cleaner, more sustainable energy paradigm has been made evident by the threat of man-made climate change (Anderegg et al., 2010; Cook et al., 2013), an uncertain stock and accessibility of fossil fuels (Heinberg & Fridley, 2010; Howarth et al., 2011; Owen et al., 2010), as well as a dependence of the energy markets on politically unstable regions of the world (Doukas et al., 2011; Hamilton, 1983). The potential contributions of IS research in reshaping energy systems have since been further outlined by, for instance, vom Brocke et al. (2013) and Goebel et al. (2014). The latter in particular point out that applied EI research has focused on energy-saving solutions and the Smart Grid. Recent publications cover a wide range of topics in this context. For instance, Loock et al. (2013) investigate behavioral aspects of the design of energy management systems. Feuerriegel et al. (2012) and Feuerriegel & Neumann (2014) analyze the costs and benefits of integrating demand response mechanisms into electricity markets. Gottwalt et al. (2011) study the demand reaction of households under flexible energy prices. The adoption of smart meters, which are generally considered an essential building block of a smart grid, is discussed in Jagstaidt et al. (2011) and Wunderlich et al. (2012). Strüker & Kerschbaum (2012) consider privacy issues associated with an increasing prevalence of smart meters. Furthermore, several research groups have proposed ways to integrate electric vehicles into a smart grid (e.g. Brandt et al., 2012; Flath et al., 2013; Kahlen et al., 2014).

Nevertheless, sustainable energy technologies still face tremendous obstacles. The adoption of electric mobility is substantially hampered by the price difference between electric and conventional vehicles (Carley et al., 2013). Similarly, the rise of renewable energy sources could only be realized with immense subsidy schemes, whose long-term consequences are difficult to predict (Kalkuhl et al., 2013). Hence, further research on building an ecologically and economically sustainable energy supply is clearly needed.

2.2 Research Approach: Design Science

Questions relating to the design, creation, and development of information systems have been central to the IS discipline since its inception (Goes, 2014). However, following Hevner et al.'s (2004) landmark article in *MIS Quarterly*, Design Science has truly established itself as a major research paradigm within the community (Land et al., 2009). Since then, the theoretical foundation of Design Science research has been further strengthened by a variety of articles, such as Hevner (2007), Gregor & Jones (2007), as well as Kuechler & Vaishnavi (2012). Nevertheless, Design Science research is still lacking presence in top-tier IS journals (Goes, 2014; Gregor & Hevner, 2013) and conferences (Bichler, 2014). This coincides with a resurgent discussion within the community concerning the general value of IS research (Hassan, 2014; Niederman, 2014).

There is a widespread sentiment that tackling questions of environmental sustainability provides a solution to both issues. On the one hand, contributing to a more sustainable way of life is perhaps the most valuable outcome any field of research can give to society (Watson et al., 2010). On the other, the relevance of Design Science in tackling environmental issues has been emphasized by, for instance, Pernici et al. (2012), Malhotra et al. (2013), and vom Brocke & Seidel (2012). In fact, many of the papers cited in the previous subsection employ a design-orientation in their approach. Hence, in this paper we seek to contribute to the Design Science knowledge base as it intersects with Energy Informatics issues. In the following subsection, we identify the specific gap in the current state of research that we address in our work.

2.3 Research Gap: IS Design in Smart Grids

The contribution of this paper is twofold. First, we closely investigate the link between IS design and the Smart Grid. Recall that a wide range of studies design artifacts for the Smart Grid with the overarching objective of improving environmental sustainability. Furthermore, it has been emphasized that Design Science is in an exceptional position to tackle sustainability challenges. Yet, little research has been done on the specific mechanisms that make IS design successful in tackling environmental issues. To address this lack of understanding, we develop a theoretical model that traces the causal chain between IS design and the realization of sustainability objectives for a broad range of applications in the Smart Grid.

The second contribution of our work lies in the case study used to underpin the theoretical reasoning. Besides its demonstrative purpose, the artifact designed in the study also serves as a practicable solution to a very real sustainability challenge. As we have outlined, the lingering adoption of green technologies stems from their high investment costs. Particularly electric vehicles, as a way to decrease the environmental footprint of the transportation sector, suffer from a substantial price difference to conventional cars. While measures to offer additional financial incentives by using electric vehicles as energy storage for the power grid have been developed, they require fleets of thousands of vehicles to function. Consequentially, they provide no solution to the immediate problem of high costs restraining customers from acquiring electric vehicles. In our case study, we present an artifact that exploits synergies between an electric vehicle and a residential photovoltaic panel. It implements smart charging strategies that increase the amount of solar energy the household itself can use. Thereby, it reduces annual energy expenses and increases the appeal of these green technologies. Furthermore, it does not require a high adoption rate of electric mobility to work—even a single household can reap its full financial benefit.

In summary, this paper outlines a specific application of IS research in the Smart Grid with the objective of improving environmental sustainability. On a broader scale, we hope that the theoretical reasoning we present results in a better understanding of how to conduct IS design research in a smart grid context to achieve specific outcomes. As we introduce the theoretical model in the next section, we will elaborate on this link between design objectives and design outcome.

3 Research Model

IS Design for Environmental Sustainability—in recent years, this catchphrase and its derivatives have been reliable guests at various IS conferences. Undoubtedly, this is a promising development, having in mind the importance of leveraging IS research to address environmental issues. We use this catchphrase as the starting point for the development of our model, since it is very unambiguous about the outcome that is to be achieved. The goal environmental sustainability implies that the impact of the conducted research should provide a tangible contribution towards a more environmentally friendly society. Such outcomes include, for instance, supporting companies in the implementation of sustainable business practices (Seidel et al., 2013), informing policymakers to improve environmental decision-making (Eickenjäger & Breitner, 2013), and using information technology to improve power supply, distribution, and consumption (Appelrath, 2012).

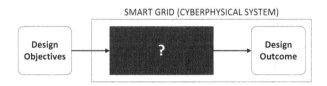

Figure II–1: Starting point: IS design for the Smart Grid

In this paper, we focus on the latter aspect—IS Design for the Smart Grid. The theoretical issue we seek to address here is that any IS design is necessarily associated with a certain anticipated outcome (cf. Figure II–1). It is the work of the researcher to translate this goal into a set of design objectives. However, the actual mechanisms in the design product that link objectives and outcome vary according to application domain. We investigate these mechanisms for IS design in a smart grid context. Considering the work of Eickenjäger & Breitner (2013) as an example, the anticipated outcome may be to empower policymakers to make informed decision concerning fossil fuels and their alternatives. The authors translate this into design objectives of, first, modeling the future development of fuel markets and, second, visualizing the results of various scenario analyses to the intended recipients. There is already a body of work on design principles that guide this type of research, most famously the guidelines by Hevner et al. (2004). However, designing for the Smart Grid poses additional challenges. Smart grids are complex cyberphysical systems—hybrids of information technology and a mix of modern as well as decades-old (legacy) power infrastructure. Successful design for the Smart Grid, thus, requires a profound understanding of the interactions between the various components in this cyberphysical system and their relationship with the people who are eventually dependent on a reliable energy supply.

Hence, our objective is to investigate the black box outlined in Figure II–1. We contemplate that one mechanism through which IS design contributes to the Smart Grid is by enhancing components of the power system with new features that were previously inaccessible. We will elaborate on this concept in the remainder of this section while we further develop the theoretical model.

3.1 Definitions

As part of our study, we discuss outcomes that are generated by information systems. Naturally, this has been extensively researched within the IS community (e.g. Markus & Silver, 2008; Robey et al., 2013; Volkoff & Strong, 2013). Since labels are not always unambiguous in the literature and we are adapting some of them slightly to fit the specifics of our context, we provide short definitions of the constructs used in our model.

Technical Object. According to Markus & Silver (2008), technical objects include IT artifacts, their components, and their output. In a Smart Grid context, this definition is somewhat problematic, since the cyberphysical system contains components of the power system that are certainly technical objects, relevant, but not necessarily IT-related. Since these two types of technical objects are distinctly different, we label them differently. During the remainder of this paper, "technical objects" exclusively refers to components of the energy infrastructure, such as power lines, generators, or batteries.

IT Artifact. This construct covers the IT-related technical objects from Markus & Silver's (2008) definition. It is the core product of the design process and can take various forms, ranging from "software, formal logic, and rigorous mathematics to informal natural language descriptions" (Hevner et al., 2004, p. 77).

Feature. Essentially, a feature is what a technical object or IT artifact does, as opposed to what it is used for. Zammuto et al. (2007) provide the example of real-time tracking sensors that indicate when a product has passed a specific stage in the process. Similarly, a feature of a battery would be that it stores energy.

Affordance. Contrary to features, affordances identify "what the user may be able to do with the object, given the user's capabilities and goals" (Markus & Silver, 2008, p. 622). This reflects the argument by Zammuto et al. (2007) that affordances arise from a combination of technological and organizational features. However, Robey et al. (2013) note that they are not limited to those the artifact's designer intended, but are shaped by human agents that use the system "with their own purposes in mind" (p. 387).

Hence, we assume that the designer conceives the IT artifact with certain features. Which affordances arise from this design and whether they achieve the intended outcome depends upon the interaction with the technical objects from the power system and the features of the socio-organizational environment.

3.2 Modeling IS Design in the Smart Grid

The Smart Grid concept comprises the reinforcement of the conventional power infrastructure with information and communication technology, as well as novel products and services arising from this hybridization. The crucial notion is that smart grids are not built from scratch, but are developed on top of an existing infrastructure. Consider, for instance, smart meters and the service environment emerging around them (Jagstaidt et al., 2011; Loock et al., 2013). They allow customers and utilities to observe the energy consumption of a particular building or device in real time. While they do grant "smartness" to the Smart Grid, they are only a tiny part of the overall cyberphysical system—the multitudes of generators, transformers, power lines, and wirings remain essentially unchanged. However, they constitute a complex system that adheres to strict laws of physics and whose stability is vital to all stakeholders.

When designing information systems for the Smart Grid, this complexity and the hybrid nature of the cyberphysical system must be kept in mind. Furthermore, the example of smart meters shows that such information systems often enhance existing infrastructure in some way. As we will show in detail in our case study, components in the power grid remain unchanged as such, but acquire new potential uses. This brings us back to Figure II–1 and the black box that contains the links between the initial design objectives and the intended outcome. As "potential uses" follow the definition of affordances by Markus & Silver (2008), we already are in the process of uncovering these links. The designer of the information system may conceive it with certain affordances in mind that would result in the intended outcome. However, these affordances arise from features of technical objects in the power system, which they acquired as a result of a certain

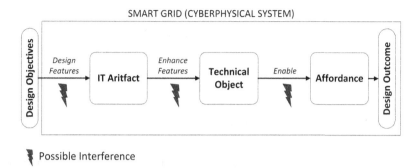

Figure II–2: Design-Interference Model for the Smart Grid

functionality of the IT artifact. It is the IS design that enhances the components of the power system with new features; their combination enables certain affordances. Whether these affordances are actualized, or whether the users employ the features for completely unanticipated uses, relies on the socio-organizational environment.

Figure II–2 visualizes our reasoning up to this point in a Design-Interference Model (DIM) for Smart Grid design products. The objectives of this model are to gain a better understanding of the relationships between the different components within the design product, but also to become aware of potential obstacles in the design process. The latter are illustrated by the *possible interferences* symbol, which draws attention to the fact that each step is a necessary prerequisite of the one following it. For instance, the enhancing feature of a technical object needs to align with the socio-organizational environment to enable the intended affordance, the logic of the IT artifact must reflect the physical restrictions on the technical object, and the design objectives have to be correctly translated into features of the IT artifact. For a successful implementation, the designer should be aware of these potential obstacles from the beginning. Hence, during the remainder of this paper we will further analyze these interferences to provide suggestions on how to conduct successful design research for the Smart Grid.

For this purpose, we proceed in the next section with a case study that provides a practical application of and validation to our model. Later, we outline possible interferences by means of the case and suggest measures to address them.

4 Case Study: Green Synergies

To further investigate the Design-Interference Model, we consider a smart grid application of IS design. The information system employs an IT artifact that uses synergies between electric mobility and residential renewable energy generation to provide additional financial incentives to households to adopt green technologies. During the course of this section, we will first review the general setting and motivation of the case. Afterwards, we outline the design of the information system, followed by an evaluation of its functionality. Throughout the section, we will draw inferences on the Design-Interference Model, which are further discussed in the subsequent section.

4.1 Setting and Motivation

The application we present seeks to address two sustainability issue society currently faces. These issues concern the increasing need for energy storage due to renewable energy sources on the one hand, and the slow adoption of electric mobility as a more sustainable means of transportation on the other. The information system designed takes advantage of synergies between renewable energy generation and electric mobility to contribute to a solution for either challenge.

To provide some background, the rising share of intermittent, decentralized, renewable energy sources requires substantial investments into distribution grids if the generation is not matched by simultaneous local demand (Nykamp et al., 2012). This emphasizes the central role of energy storage in a sustainable power system, as it detaches consumption from generation. However, storage technologies often exhibit substantial drawbacks, such as high costs (batteries), low efficiency (power-to-gas), specific geological requirements (compressed air), or a tremendous impact on landscapes and ecosystems (pumped hydro). Hence, the idea of using electric vehicles (EVs) as a swarm of decentralized storage devices has gained significant momentum. While the battery pack of an EV is quite expensive, the primary function of the vehicle is mobility. Therefore, the storage capacity can be considered an additional benefit. Kempton & Tomić (2005) have analyzed the technical feasibility of bidirectional energy flows between vehicles and the grid—so-called vehicle-to-grid (V2G) concepts. Various business models building upon these concepts have been proposed, which largely consider fleet operators and aggregators (Brandt et al., 2012; Guille & Gross, 2009; Hill et al., 2012; White & Zhang, 2011). This focus reflects the entry requirements for energy markets, which necessitate a large number of EVs, the reduction in uncertainty caused by this large number of EVs, and possibly scheduled parking times within vehicle fleets. However, the timely realization of these business approaches is questionable, since high prices hamper the public adoption of electric mobility (Al-Alawi & Bradley, 2013). The situation is quite tricky, as using electric vehicles for energy storage could provide owners with revenues that compensate for the high initial costs, yet, the realization of these revenues requires a large number of electric vehicles. To solve the immediate problem of low electric mobility adoption rates, households require incentives that compensate the high initial cost of acquiring an electric vehicle. These incentives must not be dependent on externalities, such as a high adoption rate of electric mobility within the general population.

We design an IT artifact that seeks to satisfy these requirements by focusing on households that own an electric vehicle as well as a photovoltaic installation. To reflect the impact of renewable energy feed-ins, we add the requirement that the strain on the grid, i.e. the excess feed-ins from photovoltaic generation needs to be reduced. With respect to the Design-Interference Model, the requirements constitute the anticipated design outcome, which is summarized as follows.

Financial Incentive. Users receive a financial benefit from using the system that reduces the impact of the high purchase cost of an electric vehicle.

Reduce Grid Strain. Using the system should decrease the amount of photovoltaic energy fed into the grid, as well as the energy sourced from the grid. Overall, the household should become more energy-autonomous.

Bottom-up Solution. The success of the system must not depend on the adoption rate of electric mobility in the overall population. Users must be able to capture the benefits of the information system even if they are the only adopters.

4.2 Information System Design

Existing business models that use electric vehicles as storage for intermittent renewable energy generation focus on aggregation schemes because they assume that these aggregators would enter large-scale energy markets to offer their services. A single EV can bring neither the capacity nor the reliability to compete in the energy market, which is, essentially, the aggregate energy supply. A potentially more useful approach to address intermittent *decentralized* generation is *decentralized* storage. We disaggregate the energy supply down to individual generators and consider how more of the energy produced by a specific generator can be used locally. In particular, we focus on residential households that own a rooftop photovoltaic (PV) panel in addition to a plugin-hybrid electric vehicle (PHEV). While only a share of potential EV users also owns a PV installation, encouraging those to adopt electric mobility may help to reach the critical threshold of vehicles necessary for business models that provide large-scale aggregation.

The underlying intuition behind our approach is that increasing the share of PV energy consumed locally by the household can achieve two of the anticipated design outcomes. The strain on the grid is reduced since less energy is fed into the grid and the household requires less energy from the grid, as well—it becomes overall more autonomous. Furthermore, residential households generally receive substantially less money for energy they feed into the grid than energy they source from the grid costs them. Hence, an increase of locally consumed PV energy automatically provides a financial benefit if it is matched by a decrease in energy purchased from the utility. The third outcome, the bottom-up aspect, must be included into the system development by design; i.e. at no time during the design process assumptions on the general adoption of electric mobility must be made. The outcomes are to be achieved through an information system that enables smart charging of the electric vehicle while it is plugged in at home. In this context, "smart charging" represents the opposite of "uncontrolled charging", with the latter implying that the vehicle is plugged in and immediately charged to full capacity at the maximum power level. Smart charging, on the other hand, refers to the charging (or discharging using V2G-technology) of the vehicle at times and power levels determined by the information system. The power system of the household considered in this study is outlined in detail in Section 8, along with a mathematical formulation of the associated physical constraints.

Hence, the design objectives to achieve the previously outlined outcomes are, first, the design of an IT artifact that controls charging of the electric vehicle and, second, the implementation of a charging logic that increases the share of photovoltaic energy consumed by the household. The general functionality of the information system is illustrated in Figure II–3 using the constructs from the Design-Interference Model. At this point, we want to emphasize that the DIM does not visualize the *design process*. Instead, it illustrates the *design product*, outlining what the product is supposed to do (design objectives), how the product functions (IT artifact, technical object,

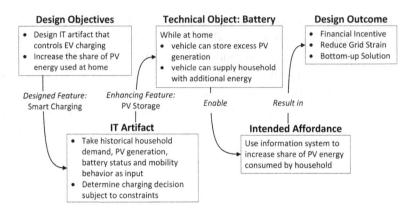

Figure II–3: Visualization of the information system functionality according to constructs of the Design-Interference Model

intended affordance), and what the product eventually delivers (design outcomes). However, by taking a step back from this idealized view of the product's functionality and considering possible interferences, the DIM allows us to draw inferences on the design process, which will be the focus of the penultimate section of this paper.

Figure II–3 points out that the design objectives are translated into an IT artifact whose primary feature is the smart charging logic, which is further illustrated in Figure II–4. In fixed intervals, the IT artifact receives information on the household demand, PV generation, the status of the PHEV battery, and historical data on the mobility behavior of the users. Based on this information, the artifact decides

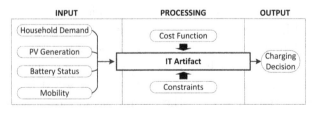

Figure II–4: Functionality of the IT artifact

whether to charge or discharge the vehicle battery (if plugged in), subject to the physical constraints of the system. Due to the high degree of uncertainty associated with the demand and the mobility behavior of a single household, the charging logic follows a threshold-based decision strategy instead of an optimization calculus. A detailed description of the implemented decision strategies follows in the next subsection and a complete mathematical formulation can be found in Section 8.

Figure II–3 further outlines that this charging logic enhances the battery of the electric vehicle. In addition to its primary purpose of providing power to the electric motor of the vehicle for traveling, the battery can buffer energy generated by the PV panel and resupply it to the household when needed. This feature—enabled by the IT artifact—allows the user to increase the share of PV energy consumed by the household. This constitutes the intended affordance of the design product and, if realized, results in the accomplishment of all anticipated design outcomes. In the final part of this section, we proceed by outlining the implemented decision strategies and by testing and evaluating the designed system using real-world data.

4.3 Testing and Evaluation

The design product is evaluated for two decision strategies in addition to the benchmark scenario (B). The latter addresses the case of uncontrolled charging, i.e. the vehicle starts charging when plugged in and stops when the battery is fully charged. The decision strategies both use a threshold for the state-of-charge of the battery up to which the vehicle is charged using excess photovoltaic energy and, if necessary, energy is procured from the grid. Above this threshold, only excess PV energy is used to further charge the battery. Conversely, if the PV generation does not suffice to satisfy the energy demand of the household, the vehicle can discharge energy to supply the household down to the threshold. The only difference between the strategies is that the *first strategy (S1) uses a fixed threshold*, while the *second strategy (S2) recalculates the threshold regularly based on the historical mobility needs of the users*. It should be noted that the influence of the threshold is not negligible. If a threshold is chosen that is too low, the energy stored in the battery may not suffice to finish a trip. Since we only consider plugin-hybrid EVs, whose combustion engines can provide the missing range, mobility is not impaired. However the high price of gasoline compared to electrical energy would substantially decrease the financial benefit of the system to the user. If the chosen threshold is too high, the amount of PV energy that can be buffered is limited, which limits the potential financial benefit, as well.

As this is a conceptual study and V2G-capable vehicles are not yet in series production, we cannot evaluate our artifact using a real-world prototype (Lee et al., 2008) or field study (Pries-Heje & Baskerville, 2008). However, V2G technology is well understood and we can use real-world data on the physical constraints to construct a realistic simulation in Matlab. Naturally, the form of the IT artifact in a prototype study is different from the version implemented in the simulation. While the latter is a Matlab script to work with the simulation, the code and functionality can be transformed into, for instance, a smartphone app in a relatively straightforward manner.

The parameters used in the evaluation are summarized in Table II–1. We consider Germany for this case study whose retail energy prices are based on data from the German Association of Energy and Water Industries (bdew, 2011). To provide general validity of our results, we correct the prices and revenues for the extensive subsidy scheme currently in place in Germany. We derive the specifications of the PHEV battery from the Chevy Volt and base the price of energy from gasoline on the average price for Germany in 2011. We use very conservative values for the maximum charging and discharging rates of the EV as the explicit objective of our research is to provide a system that is beneficial without extensive technological requirements. Hence, the maximum charging rate relates to a standard residential 220V/240V outlet and the maximum discharging rate (V2G) is derived

Table II–1: Evaluation parameters

Parameter	Value
Retail price of energy	0.2142 EUR per kWh
Revenue from PV energy fed into grid	0.0500 EUR per kWh
Price of PV energy consumed by household	0.0000 EUR per kWh
Price of energy from gasoline	0.4408 EUR per kWh
Minimum energy stored in battery	0 kWh
Maximum energy stored in battery	16 kWh
Maximum charging power of battery	800 W
Maximum discharging power of battery	500 W
Losses from inversion and other causes	0.93

from the values provided by Mitsubishi for its i-MiEV electric car. The parameter on loss from inversion is taken from Kempton & Tomić (2005) as this value is still valid for most common inverters. Furthermore, we use an original trace of PV generation, taken from a PV installation on a single-family home in Eastern Bavaria throughout 2011. This trace describes PV generation in 5-minute intervals for the entire year. This interval size is subsequently used for all calculations. The demand trace is from a specific German 4-person household in 2011.

Finally, we evaluate the performance of the information system for three distinct types of users, which are distinguished by their mobility needs. The first type (T1) is an employee working fixed shifts, implying a reliable, repetitive mobility profile. The second type (T2) is a family with fixed schedules on weekdays. The profile is quite repetitive as well, but trips are undertaken at different times and are of different lengths than for the shift worker. The third profile (T3) is, again, for a family, but the trips undertaken are randomized based on mobility profiles taken from the study *Mobility in Germany, 2008* (BMBVS, 2010), which provides a very comprehensive dataset on the mobility

Table II–2: Annual energy costs and annual effect on the grid

Type	Strategy	Energy Cost	PV Energy Feed-In	Energy Purchase
	B	EUR 467.60	4,881.05 kWh	3,322.38 kWh
T1	S1	EUR 271.87	3,571.16 kWh	2,102.84 kWh
	S2	EUR 249.52	3,364.84 kWh	1,950.36 kWh
	B	EUR 701.60	4,254.90 kWh	4,268.63 kWh
T2	S1	EUR 493.19	2,828.83 kWh	2,962.78 kWh
	S2	EUR 394.94	2,297.04 kWh	2,360.63 kWh
	B	EUR 410.57	4,957.91 kWh	3,074.06 kWh
T3	S1	EUR 210.86	3,565.78 kWh	1,816.75 kWh
	S2	EUR 152.58	3,067.94 kWh	1,428.11 kWh

The evaluation is conducted for households featuring three different mobility types over 1 year. T1 and T2 feature a shift-worker and a family, respectively, with fixed recurring mobility patterns. T3 features a family with one working parent and the profile is randomized based on real-worl data. The strategies evaluated are benchmark (B), decision-strategy with fixed threshold (S1), and decision-strategy with flexible threshold (S2).

needs of the German population. This profile is certainly the most realistic one, but overall the results from all three types should provide indication and validation of the information system's performance.

Table II–2 provides an overview of the evaluation results for the various household types and strategies. We can observe that for any household type the decision strategies substantially outperform the benchmark values. Furthermore, the strategy with flexible thresholds (S2) consistently performs best, although the difference to the fixed threshold ranges from about 22 euros for the shift-worker to 100 euros for the family. The most realistic randomized profile results in additional cost decrease of about 60 euros over S1.

If we consider the total cost reduction for T3, the household can decrease annual energy costs by 258 euros compared to the benchmark. These reductions equal about two to three monthly electricity bills for the average German 4-person household every year. Compare this to the federal subsidy for plugin-hybrid EVs in the United States following the American Recovery and Reinvestment Act of 2009. The US government subsidizes a Chevy Volt with a tax credit of USD 7,500, which equals about EUR 5,800 given current exchange rates. At a marginal tax rate of 30 percent, the actual financial incentive to households is EUR 1,740. This equals about six to seven years of running the information system designed in this study, which is easily within the lifetime of an EV battery. So, adopting the information system can certainly provide a noteworthy financial incentive for electric mobility. However, the results in Table II–2 also outline that the households become substantially more autonomous. The amount of excess PV energy fed into the grid, as well as the amount of energy procured from the grid, are substantially reduced. This decreases the overall strain on the distribution grid for every household adopting the system.

To conclude this evaluation of the information system designed in this case study, Figure II–5 outlines how the strategies change the composition of the energy supplied to the household for type T3. The benchmark only uses PV and grid supply, but we can observe that PV supply exceeds household demand several times per day (e.g. between 6.30 a.m. and 8.30 a.m.). Particularly during morning hour,s this energy is often fed into the grid since the PHEV is still at full battery. This is already partially compensated when using strategy S1 with a fixed threshold. The PHEV supplies most of the missing energy during the day, only when the battery threshold is reached is the remaining energy sourced from the grid. Panel c in Figure II–5 shows the composition of energy supply for strategy S2. The flexible battery threshold allows eliminating almost any need for energy from the grid during this particular day. For both V2G strategies, the energy the PHEV battery supplies to the household is then recharged by using

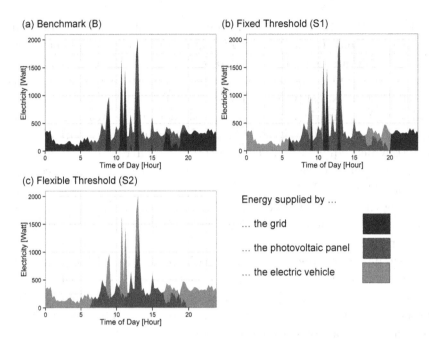

Figure II–5: Composition of energy supply to household for different
            strategies

excess PV energy or, if necessary, energy from the grid the following
day.

Overall, the IT artifact uses information on the user's driving behavior
to enhance the battery of the electric vehicle with the feature of
buffering photovoltaic energy for later use by the household. Thus, it
makes existing synergies between these technologies usable.

5 From Objectives to Outcomes: Preventing Interferences

In this section we return to the Design-Interference Model to derive implications on successful IS research for the Smart Grid from the case study we have discussed. As visualized in Figure II–2, possible interferences are most likely at the points of contact between the constructs of the model. Consequently, we discuss these points of interference and the lessons learned for future research endeavors.

5.1 Point of Interference: From Design Objectives to the IT Artifact

The design process is instantiated by the transformation of the design objectives into a tangible IT artifact. The IS designer defines the eventual design product and considerations of all possible interferences must be incorporated, be they in shaping the artifact, in the relationship with the technical system, or in the behavioral and organizational environment. Interferences at this point are common for all applications of IS design and not limited to a smart grid context. They range from abstract, such as flaws in the general architecture of the artifact, to practical, such as coding errors. Guidelines on combating these interferences are outlined in Hevner et al. (2004). While they are applicable to all types of IS design research, we perceive the following three of them to be of particular relevance when designing information systems for the Smart Grid. Based on our case study, we will propose three additional guidelines specifically for smart grid research later in this section.

> *Design as an Artifact.* The importance of this guideline is evident as we have been discussing IT artifacts throughout this paper. The reasons are twofold. First, sustainability in general and the Smart Grid in particular are transdisciplinary and societally relevant issues. An IT artifact

provides a tangible research result that can reach a large target audience in the scientific community, as well as the general population, without extensive requirements on their educational background. Second, energy is ubiquitous, yet invisible; it is crucial to the daily routines of billions of people and the operations of countless companies. Designing as an artifact enables researches to better communicate how the research results will affect the daily personal and corporate "energy experiences". For instance, in the context of our case study, we can present the designed artifact as an application that decreases annual energy costs without impeding mobility or daily activities.

Design Evaluation. As we have outlined, smart grids are complex systems and in certain ways more complex than many other applications of IS design. This results from the inclusion of components of the cyberphysical system that may be unfamiliar to the IS researcher. Even in interdisciplinary research teams, this may lead to unexpected performance deviations of the design product. Rigorous evaluation is indispensable and should be executed at multiple stages of the design process, if practicable. Successful examples range from evaluating the theoretical conception through a mathematical formulation of the expected outcomes, and evaluating pre-implementation through simulation using real-world data as conducted in our case study, to an evaluation post-implementation using field studies.

Design as a Search Process. The complexity of smart grid research causes many solutions to not be immediately, or even after some consideration, evident. For instance, it is generally assumed that individual electric vehicles cannot effectively provide energy storage without some measure of aggregation. In our case study, we show that this can be achieved by scaling down the target system and by foregoing an optimization calculus for threshold-based decision strategies. However, finding this kind of solutions requires letting go of preconceived notions and searching for what works instead of what one wants to work.

5.2 Point of Interference: From the IT Artifact to the Technical Object

The point of interference between the IT artifact and the technical object of the legacy system is specific for information systems in cyberphysical systems, such as the Smart Grid. We want to outline two major sources of interferences in a smart grid context and guidelines on how to avoid them. First, the physical system might not interact with and react to the IT artifact as expected. For instance, in our case study we assume that the EV battery is fully dischargeable. However this kind of deep discharge may result in a much shorter lifespan of the battery. We omitted such lifespan issues in our case study simply because in our scenarios the charging strategies never caused additional deep charging incidents and an inclusion of the effects on the battery would unnecessarily complicate the paper. Nonetheless, it is important that we are aware of the potential issues, leading to our first additional guideline.

Comprehend the System. It is vital that the legacy system is well understood by the researchers. This suggestion may appear obvious or even trivial at first, but it is so crucial to successful design of information systems for smart grids that it must be mentioned. Questions of sustainability and the Smart Grid as an aspect of this broad field are inherently interdisciplinary and require knowledge from various fields to tackle. While successful IS design for smart grids does not necessarily require interdisciplinary teams, the researchers involved should have a thorough understanding of the technical and behavioral aspects of smart grids and energy. Furthermore, for this research to have an impact beyond the IS discipline, this understanding must be communicated in the articles and publication outlets must allow researchers space to communicate it in some form in their papers. Lastly, comprehension of the system is necessary to find synergies between different technologies within the system, to design information systems that enhance existing devices, and allow them to provide more benefits.

The second source of interferences, but also of opportunities, at the link between the IT artifact and the technical object results from the fact that power systems are networks. What works for an individual application, might end up destabilizing the system if it is adopted by many users. On the other hand, the full benefit of a certain IT artifact may only be realizable once many users adopt it. This issue is articulated in our second guideline.

Consider Network Effects. Designers must be aware of the impact, as well as the attainability, of network effects. Consider the research undertaken in our case study. The motivation was network effects—the ability of many electric vehicles to serve as energy storage for intermittent energy generation from renewable sources. However, what further guided our search process was the fact that these network effects are currently unattainable, due to the lacking adoption rates of electric mobility. *Green Synergies*, i.e. the interaction between various green technologies such as photovoltaics and electric mobility, are also a kind of network effect, although not in the most classical sense. Nevertheless, they arise from a holistic perspective on the complete system, which is indispensable for successful IS design in smart grids.

5.3 Point of Interference: From Enhancing Features to Enhancing Affordances

The step from feature to affordance is a core subject of Information Systems research and fully discussing it in light of our model and case study would easily fill another paper. Questions, such as "How to get the user to use the artifact for the intended purpose?" and "Which potential uses materialize that the designer did not anticipate, for good or for bad?" are of crucial importance to IS research of any kind. Unintended affordances that are frequently discussed in the context of smart grids relate to privacy issues since smart meters may open the gates to potential intruders and abusers. However, privacy issues are also relevant to any information system implementation. Instead, we want to use the final guideline, this final note in our discussion on an obvious feature of the information system in our case study at the intersection between technical system and the resulting affordances—the role of the user is marginal. In fact, information on all relevant behavior is collected by sensors, such as smart meters or the IT of the electric vehicle. Essentially, the user only has to acquire the system and its operations are completely automated. Hence, some might question the fit of this system with the general orientation of IS research. We disagree, since minimizing the reliance on the user was a conscious decision on part of the designers and the role and the influence of the user are questions that must be asked in IS design in general, but in a system as complex as a smart grid in particular. Yet, who is better equipped to tackle this question other than IS researchers?

Without doubt, there is potential how a more explicit user inter-action with the system could improve the results and increase the financial benefits as well as the energy autonomy of the household. If users supply information on future trips, the critical threshold of the battery can be tuned to finer levels, maximizing the enhancing feature of the battery to buffer photovoltaic energy. However, this potential improvement must be considered in light of the probable price. First, increasing user interaction requires a user interface, which increases development time and costs, as well as potentially the cost of the system to the end-user. Second, we need to ask whether the user *can* provide information that will substantially improve the outcomes. Trips are often undertaken spontaneously and it is unclear if an optimization based on incorrect predictions results in an actual improvement. Third, we need to ask whether the user *is willing* to provide the necessary information, whether he will use the system in the way the designer intended. While the overall financial benefit of the system is quite large, the daily revenues are around one euro. It is questionable whether this is a sufficient incentive to urge users to report their future trips to the system. The careful consideration of these facts led us to minimizing the role of the user in our case study. While this implies by no means that the user must always be marginalized in smart grid information systems, it leads to our final guideline.

> *Question the Role of the User.* A substantial user interaction might substantially increase the performance of the system. It might also be necessary by design. It might also open the gates to detrimental, unintended affordances arising from the system. The complexity of the cyberphysical system the user may have to deal with could be overwhelming. There is no general rule on how much user interaction with the technical system is optimal, but it is important to be open-minded on the one hand and critical on the other hand.

6 Conclusion

In this paper, we have demonstrated the potential of IS research for the Smart Grid as a part of the overarching goal of improving environmental sustainability. We have introduced the Design-Interference Model for the Smart Grid, which analyzes the design product along with its antecedents (the design objectives) and impact (the design outcomes). Furthermore, it provides insights on the sources of possible interferences in the design process. We propose that one central way how information systems contribute to smart grids is by enhancing components of the existing legacy power infrastructure. We substantiate this claim using a case study of an information system that enhances the battery of an electric vehicle. However, the concept of enhancing features can also be observed in other applications of smart metering and smart grids (e.g. Katz et al., 2011).

By analyzing the case study in light of the Design-Interference Model, we are able to derive a set of guidelines for IS research in a smart grid context, which expand on the work by Hevner et al. (2004) on general Design Science projects. As illustrated in Figure II–6, we first emphasize the relevance of designing as an artifact, design evaluation, and design as a search process for smart grid research. Furthermore, to prevent possible interferences associated with the cyberphysical nature of the smart grid, we propose three additional guidelines of comprehending the system, considering network effects, and questioning the role of the user.

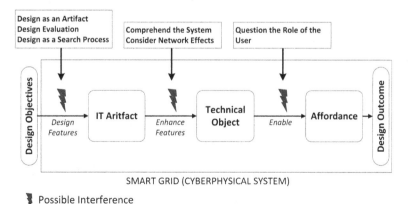

Figure II–6: Design-Interference Model with guidelines to prevent interferences

Through our case study, we could demonstrate the tremendous beneficial impact IS design can have on future, more sustainable power systems. We hope that this example encourages scholars to undertake this challenging but rewarding work and that our guidelines support them in this process.

7 References

Al-Alawi, B. M. & Bradley, T. H. (2013). Total cost of ownership, payback, and consumer preference modeling of plug-in hybrid electric vehicles. *Applied Energy, 103*, 488–506.

Anderegg, W. R. L., Prall, J. W., Harold, J. & Schneider, S. H. (2010). Expert credibility in climate change. *Proceedings of the National Academy of Sciences of the United States of America, 107*(27), 12107–12109.

Appelrath, H.-J. (2012). Interview with Henning Kagermann on "Smart Grids – Information and Communication Technology as a Key Factor in a Future Energy System". *Business & Information Systems Engineering, 4*(1), 45–46.

bdew. (2011). *Costs of renewable energies continue to rise moderately*. Berlin. Retrieved on April 13, 2015, from `http://goo.gl/mpJHiX`

Bichler, M. (2014). Reflections on the State of Design Science Research. *Business & Information Systems Engineering, 6*(2), 71–72.

BMBVS. (2010). *Mobility in Germany, 2008*. Bonn, Berlin: Federal Ministry of Transport, Building and Urban Development.

Brandt, T., Wagner, S. & Neumann, D. (2012). Road to 2020: IS-Supported Business Models for Electric Mobility and Electrical Energy Markets. *ICIS 2012 Proceedings*, Paper 48.

Carley, S., Krause, R. M., Lane, B. W. & Graham, J. D. (2013). Intent to purchase a plug-in electric vehicle: A survey of early impressions in large US cites. *Transportation Research Part D: Transport and Environment*, *18*, 39–45.

Cook, J., Nuccitelli, D., Green, S. A., Richardson, M., Winkler, B., Painting, R., ... Skuce, A. (2013). Quantifying the consensus on anthropogenic global warming in the scientific literature. *Environmental Research Letters*, *8*(2), 024024.

Corbett, J. (2013). Designing and Using Carbon Management Systems to Promote Ecologically Responsible Behaviors. *Journal of the Association for Information Systems*, *14*(7), 339–378.

Doukas, H., Flamos, A. & Psarras, J. (2011). Risks on the Security of Oil and Gas Supply. *Energy Sources, Part B: Economics, Planning, and Policy*, *6*(4), 417–425.

Eickenjäger, M.-I. & Breitner, M. (2013). REFUSA: IS-Enabled Political Decision Support with Scenario Analyses for the Substitution of Fossil Fuels. *ICIS 2013 Proceedings*, Paper 5.

Elliot, S. (2011). Transdisciplinary Perspectives on Environmental Sustainability: A Resource Base and Framework for IT-Enabled Business Transformation. *Management Information Systems Quarterly*, *35*(1), 197–236.

Farhangi, H. (2010). The path of the smart grid. *IEEE Power and Energy Magazine*, *8*(1), 18–28.

Feuerriegel, S. & Neumann, D. (2014). Measuring the financial impact of demand response for electricity retailers. *Energy Policy*, *65*, 359–368.

Feuerriegel, S., Strüker, J. & Neumann, D. (2012). Reducing Price Uncertainty through Demand Side Management. *ICIS 2012 Proceedings*, Paper 7.

Flath, C. M., Ilg, J. P., Gottwalt, S., Schmeck, H. & Weinhardt, C. (2013). Improving Electric Vehicle Charging Coordination Through Area Pricing. *Transportation Science, 48*(4), 619–634.

Goebel, C., Jacobsen, H.-A., Razo, V., Doblander, C., Rivera, J., Ilg, J., ... Lässig, J. (2014). Energy Informatics. *Business & Information Systems Engineering, 6*(1), 25–31.

Goes, P. (2014). Editor's Comments: Design Science Research in Top Information Systems Journals. *Management Information Systems Quarterly, 38*(1), iii–viii.

Gottwalt, S., Ketter, W., Block, C., Collins, J. & Weinhardt, C. (2011). Demand side management—A simulation of household behavior under variable prices. *Energy Policy, 39*(12), 8163–8174.

Gregor, S. & Hevner, A. (2013). Positioning and Presenting Design Science Research for Maximum Impact. *Management Information Systems Quarterly, 37*(2), 337–355.

Gregor, S. & Jones, D. (2007). The Anatomy of a Design Theory. *Journal of the Association for Information Systems, 8*(5), Article 2.

Guille, C. & Gross, G. (2009). A conceptual framework for the vehicle-to-grid (V2G) implementation. *Energy Policy, 37*(11), 4379–4390.

Hamilton, J. D. (1983). Oil and the Macroeconomy since World War II. *Journal of Political Economy, 91*(2), 228–248.

Hassan, N. (2014). Value of IS Research: Is there a Crisis? *Communications of the Association for Information Systems, 34*(1), Article 41.

Heinberg, R. & Fridley, D. (2010). The end of cheap coal. *Nature, 468*(7322), 367–369.

Hevner, A. (2007). A Three Cycle View of Design Science Research. *Scandinavian Journal of Information Systems, 19*(2), Article 4.

Hevner, A., March, S., Park, J. & Ram, S. (2004). Design Science in Information Systems Research. *Management Information Systems Quarterly*, *28*(1), 75–106.

Hill, D. M., Agarwal, A. S. & Ayello, F. (2012). Fleet operator risks for using fleets for V2G regulation. *Energy Policy*, *41*, 221–231.

Hilpert, H., Schumann, M. & Kranz, J. (2013). Leveraging Green IS in Logistics. *Business & Information Systems Engineering*, *5*(5), 315–325.

Howarth, R. W., Ingraffea, A. & Engelder, T. (2011). Natural gas: Should fracking stop? *Nature*, *477*(7364), 271–275.

Jagstaidt, U., Kossahl, J. & Kolbe, L. (2011). SmartMetering Information Management. *Business & Information Systems Engineering*, *3*(5), 323–326.

Kahlen, M., Ketter, W. & van Dalen, J. (2014). Balancing with Electric Vehicles: A Profitable Business Model. *ECIS 2014 Proceedings*, Paper 11.

Kalkuhl, M., Edenhofer, O. & Lessmann, K. (2013). Renewable energy subsidies: Second-best policy or fatal aberration for mitigation? *Resource and Energy Economics*, *35*(3), 217–234.

Katz, R. H., Culler, D. E., Sanders, S., Alspaugh, S., Chen, Y., Dawson-Haggerty, S., ... Shankar, S. (2011). An information-centric energy infrastructure: The Berkeley view. *Sustainable Computing: Informatics and Systems*, *1*(1), 7–22.

Kempton, W. & Tomić, J. (2005). Vehicle-to-grid power fundamentals: Calculating capacity and net revenue. *Journal of Power Sources*, *144*(1), 268–279.

Kuechler, W. & Vaishnavi, V. (2012). A Framework for Theory Development in Design Science Research: Multiple Perspectives. *Journal of the Association for Information Systems*, *13*(6), 395–423.

Land, F., Loebbecke, C., Angehrn, A., Clemons, E., Hevner, A. & Mueller, G. (2009). ICIS 2008 Panel Report: Design Science in Information Systems: Hegemony, Bandwagon, or New Wave? *Communications of the Association for Information Systems*, *24*(1), Article 29.

Lee, J., Wyner, G. & Pentland, B. (2008). Process Grammar as a Tool for Business Process Design. *Management Information Systems Quarterly*, *32*(4), 757–778.

Loock, C.-M., Staake, T. & Thiesse, F. (2013). Motivating Energy-Efficient Behavior with Green IS: An Investigation of Goal Setting and the Role of Defaults. *Management Information Systems Quarterly*, *37*(4), 1313–1332.

Malhotra, A., Melville, N. & Watson, R. T. (2013). Spurring Impactful Research on Information Systems for Environmental Sustainability. *Management Information Systems Quarterly*, *37*(4), 1265–1274.

Markus, M. L. & Silver, M. (2008). A Foundation for the Study of IT Effects: A New Look at DeSanctis and Poole's Concepts of Structural Features and Spirit. *Journal of the Association for Information Systems*, *9*(10), Article 5.

Melville, N. (2010). Information Systems Innovation for Environmental Sustainability. *Management Information Systems Quarterly*, *34*(1), 1–21.

Niederman, F. (2014). Responding to Three Issues in Hassan (2014). *Communications of the Association for Information Systems*, *34*(1), Article 44.

Nykamp, S., Molderink, A., Hurink, J. L. & Smit, G. J. (2012). Statistics for PV, wind and biomass generators and their impact on distribution grid planning. *Energy*, *45*(1), 924–932.

Owen, N. A., Inderwildi, O. R. & King, D. A. (2010). The status of conventional world oil reserves—Hype or cause for concern? *Energy Policy*, *38*(8), 4743–4749.

Pernici, B., Aiello, M., vom Brocke, J., Donnellan, B., Gelenbe, E. & Kretsis, M. (2012). What IS Can Do for Environmental Sustainability: A Report from CAiSE'11 Panel on Green and Sustainable IS. *Communications of the Association for Information Systems*, *30*(1), Article 18.

Pries-Heje, J. & Baskerville, R. L. (2008). The Design Theory Nexus. *Management Information Systems Quarterly*, *32*(4), 731–755.

Robey, D., Anderson, C. & Raymond, B. (2013). Information Technology, Materiality, and Organizational Change: A Professional Odyssey. *Journal of the Association for Information Systems*, *14*(7), Article 1.

Seidel, S., Recker, J., Pimmer, C. & vom Brocke, J. (2014). IT-enabled Sustainability Transformation—the Case of SAP. *Communications of the Association for Information Systems*, *35*(1), Article 1.

Seidel, S., Recker, J. & vom Brocke, J. (2013). Sensemaking and Sustainable Practicing: Functional Affordances of Information Systems in Green Transformations. *Management Information Systems Quarterly*, *37*(4), 1275–1299.

Strüker, J. & Kerschbaum, F. (2012). From a Barrier to a Bridge: Data-Privacy in Deregulated Smart Grids. *ICIS 2012 Proceedings*, Paper 2.

Volkoff, O. & Strong, D. (2013). Critical Realism and Affordances: Theorizing IT-Associated Organizational Change Processes. *Management Information Systems Quarterly*, *37*(3), 819–834.

vom Brocke, J. & Seidel, S. (2012). Environmental Sustainability in Design Science Research: Direct and Indirect Effects of Design Artifacts. In D. Hutchison et al. (Eds.), *Design Science Re-*

search in Information Systems. Advances in Theory and Practice (Vol. 7286, pp. 294–308). Berlin, Heidelberg: Springer Berlin Heidelberg.

vom Brocke, J., Watson, R., Dwyer, C., Elliot, S. & Melville, N. (2013). Green Information Systems: Directives for the IS Discipline. *Communications of the Association for Information Systems*, *33*(1), Article 30.

Watson, R., Boudreau, M.-C. & Chen, A. (2010). Information Systems and Environmentally Sustainable Development: Energy Informatics and New Directions for the IS Community. *Management Information Systems Quarterly*, *34*(1), 23–38.

White, C. D. & Zhang, K. M. (2011). Using vehicle-to-grid technology for frequency regulation and peak-load reduction. *Journal of Power Sources*, *196*(8), 3972–3980.

Wunderlich, P., Veit, D. & Sarker, S. (2012). Examination of the Determinants of Smart Meter Adoption: An User Perspective. *ICIS 2012 Proceedings*, Paper 13.

Zammuto, R. F., Griffith, T. L., Majchrzak, A., Dougherty, D. J. & Faraj, S. (2007). Information Technology and the Changing Fabric of Organization. *Organization Science*, *18*(5), 749–762.

8 Appendix

8.1 Physical System

Figure II–7 illustrates the components of and the possible energy flows within the physical system. The information system determines its charging decisions using 5-minute intervals. The household and the mobility needs are pure consumers while the photovoltaic panel and the hybrid motor of the vehicle are pure suppliers of energy. The

Table II–3: Constraints on physical system

The household demand must be satisfied, either by PV energy, grid energy, or energy supplied by the PHEV if it is plugged in. $$D_t = PD_t + GD_t + \beta_t BD_t$$
All energy generated by the PV installation must be distributed somewhere, either the household, the grid, or the plugged-in PHEV. $$P_t = PD_t + PG_t + \beta_t PB_t$$
The net position of the grid is energy supplied to household and PHEV minus the energy received from the PV panel. $$G_t = GD_t + \beta_t GB_t - PG_t$$
The change of the state of charge of the battery within period t is either energy supplied by the grid or the PV panel minus the energy supplied to the household (if the vehicle is plugged in), or the energy supplied by the hybrid motor minus the energy required for mobility (if the vehicle is not plugged in). $$B_t - B_{t-1} = \beta_t \left[\alpha PB_t + \alpha GB_t - \tfrac{1}{\alpha} BD_t \right] + (1 - \beta_t) \left[HB_t - BM_t \right]$$
Mobility needs must be satisfied. $$M_t = BM_t$$
All energy from the hybrid motor goes to the battery. $$H_t = HB_t$$
The variables are limited to the following ranges. In addition, the state of charge of the battery must be within the range defined by the minimum and the maximum state of charge. Charging and discharging of the battery are limited by the maximum charge and discharge rates, respectively. $$D_t, P_t, M_t, H_t, PD_t, GD_t, BD_t, PG_t, PB_t, GB_t, HB_t, BM_t \geq 0,$$ $$0 < \alpha \leq 1, \quad \beta_t \in {0,1}, \quad B_{MIN} \leq B_t \leq B_{MAX},$$ $$\tfrac{1}{\alpha} BD_t \leq B_{OUT}, \quad \alpha(PB_t + GB_t) \leq B_{IN}$$

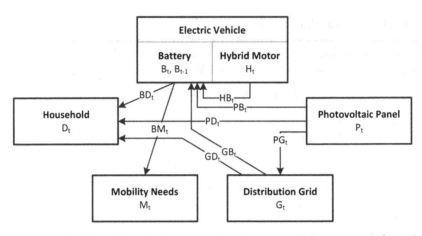

D$_t$ Household demand in period t   B$_t$ Energy stored in **b**attery at end of period t
M$_t$ Energy required for **m**obility in period t   H$_t$ Energy supplied by **h**ybrid motor in period t
G$_t$ **G**rid supply or demand in period t
P$_t$ **P**hotovoltaic generation in period t   XY$_t$ Energy flow *from* **X** *to* **Y** in period t

Figure II–7: Schematic representation of physical system

battery and the distribution grid can consume or supply energy within their physical limits.

Table II–3 outlines the resulting constraints on the physical system, which have to be satisfied in each period t. The parameter β indicates whether the vehicle is plugged in at home (1) or not (0). The parameter α represents losses from inversion.

Finally, the total cost of energy in period t is calculated as

$$C_t = p_r(GB_t + GD_t) + p_f PG_t + p_c(PB_t + PD_t) + p_g HB_t$$

with p_r as the retail price of energy, p_f as the price of PV energy fed into the grid (generally negative, as it is a revenue of the household), p_c as the price of PV energy consumed by the household (equal to zero in the case study, may be negative in certain subsidy schemes), and p_g as the price of gasoline.

8.2 Implemented Strategies

The additional constrains that must be satisfied for the benchmark are outlined in Table II–4 while the constraints for the V2G strategies are listed in Table II–5. As previously mentioned, both strategies follow the same logic, their only difference lies in the calculation of the threshold state of charge of the battery, B_{crit}. While this value is fixed for S1, it is permanently recalculated for S2 based on the historical mobility needs of the users.

Table II–4: Constraints for benchmark (B)

There is no energy flow from the vehicle to the household.
$$BD_t = 0$$
The first priority of PV generation is to satisfy household demand.
$$PD_t = \min[P_t, D_t]$$
The remaining PV energy is used to charge the PHEV, if plugged in, until the battery is fully charged.
$$PB_t = \beta_t \min[\max[P_t - PD_t, 0], B_{MAX} - B_{t\ 1}]$$
If plugged in, the PHEV is charged with maximum power.
$$PB_t + GB_t = \beta_t \min[B_{IN}, B_{MAX} - B_{t\ 1}]$$

Table II–5: Constraints for decision strategies (S1 and S2)

The first priority of PV generation is to satisfy household demand.
$$PD_t = \min[P_t, D_t]$$
If plugged in and battery level is below critical threshold, the PHEV is charged with maximum power. Otherwise, only excess PV energy is used until the battery is fully charged.
$$PB_t + GB_t = \begin{cases} \beta_t B_{IN} & \text{if } B_{t\ 1} \leq B_{crit} \\ \beta_t \min[\max[P_t - PD_t, 0], B_{MAX} - B_{t\ 1}] & \text{otherwise} \end{cases}$$
If plugged in and battery level is above critical threshold, remaining household demand is supplied by EV until either maximum power or critical threshold is reached.
$$BD_t = \begin{cases} \beta_t \min[D_t - PD_t, B_{t\ 1} - B_{crit}, B_{OUT}] & \text{if } B_{t\ 1} > B_{crit} \\ 0 & \text{otherwise} \end{cases}$$

III Designing an Energy Information System for Microgrid Operation

This chapter is a revised version of the paper:

Tobias Brandt, Nicholas DeForest, Michael Stadler, and Dirk Neumann, "Power Systems 2.0: Designing an Energy Information System for Microgrid Operation" (2014), ICIS 2014 Proceedings, Paper 8.

Part of this work has been financed by the Office of Electricity Delivery and Energy Reliability, Distributed Energy Program of the U.S. Department of Energy under Contract No. DE-AC02-05CH11231.

Abstract

In this paper we demonstrate the contribution of information systems towards a sustainable and reliable power supply. Following a Design Science approach, we develop an information system for microgrid operation at a U.S. army base. Microgrids enable an improved integration of distributed renewable energy sources and increase the robustness of the overall power grid. The microgrid in this study contains extensive photovoltaic generation, the energy demand of the base, as well as energy storage. The information system we design controls and optimizes microgrid operations under uncertainty, as well as physical and organizational constraints. Using real-world data to evaluate the system, we show that it substantially increases the amount of clean photovoltaic energy that can be generated while simultaneously decreasing energy costs of the base. Thereby, we are able to improve the ecological and economic efficiency of the microgrid.

1 Introduction

As part of the discussion on how information systems can help society shape a more prosperous future and a sustainable way of living, the concept of *Energy Informatics* (EI) has received widespread attention in recent years (Goebel et al., 2014). A core concern of EI research has been the changing shape of the power grid, with *information and communication technology* (ICT) complementing traditional power systems. An emphatic push of societies around the globe away from fossil fuels towards renewable energy sources holds the promise of a cleaner energy supply but also comes with new challenges that need to be overcome. Decentralized energy generators, such as photovoltaic panels or *combined heat and power* (CHP) units, pose challenges to grid operators and the traditional grid structure (Paatero & Lund, 2007). Similarly, the volatility of some renewable energy sources requires novel demand side management measures that employ ICT infrastructure to align supply and demand in the grid (Strueker & Dinther, 2012).

In this paper we present a case study that analyzes how *microgrids* can alleviate these issues and we show the substantial contribution IS research can make towards the implementation of microgrid concepts. Microgrids are localized electricity distribution systems that operate in a controlled, coordinated way, either while connected to the main power network or while islanded (R. H. Lasseter & Paigi, 2004). Essentially, microgrids attempt to locally balance supply and demand, thereby reducing the volatility of the overall system. In addition to an improved integration of renewable energy sources, they also provide a more resilient power infrastructure in the face of disaster (Kwasinski et al., 2012) and new opportunities for establishing rural electrification in third-world countries (Chaurey & Kandpal, 2010; Nandi & Ghosh, 2009). The concept of microgrids has been researched for several years with respect to their physical foundations, as well as their theoretical economic feasibility. The role of IS research becomes evident when considering the implementation of a microgrid and the associated cost-efficient management of supply and demand, given fluctuating loads and generation, flexible energy tariffs, as well as organizational constraints.

The case study discussed in this research concerns a microgrid to be implemented at a U.S. army base. The base contains a two megawatts photovoltaic (PV) installation (which is to be increased by an additional megawatt). Since only a maximum of one megawatt of power may be exported at any time, parts of the PV installation frequently need to be disconnected from the grid, essentially wasting energy that could potentially have been generated. As this contradicts the government's stated environmental goals, a one megawatt-hour battery has been installed. Adhering to the design science research paradigm (e.g. Hevner et al., 2004), we design an IS artifact that determines operational decisions for the microgrid given the mentioned, as well as other physical, economic, and organizational restrictions. Specifically, we address the following research questions:

RQ1: What requirements does the information system need to satisfy? (*Requirements Analysis*)

RQ2: How does the corresponding information system need to be designed? (*Artifact Design*)

RQ3: Are the initial requirements satisfied and what is estimated benefit of the IS-enhanced microgrid? (*Evaluation*)

While the IS artifact consists of several modules, the optimization module employs DER-CAM, an optimization tool for microgrid operation developed at Berkeley National Laboratory. DER-CAM will be introduced in more detail in the following section, along with related work from the IS community, as well as from other disciplines that have conducted microgrid research. The third section provides an overview of the research design of the entire microgrid project and defines the scope of this study in its context. The requirements analysis (RQ1) is also addressed in this section. The subsequent section contains the actual design process and the evaluation of the resulting artifact (RQ2 and RQ3), followed by a section discussing the implications of our work. We conclude by summarizing this study and pointing out possible paths for future research.

2 Relevance to IS Research and Related Work

The research questions that are raised in this study address the issue of how information systems can increase the effectiveness and efficiency of modern power systems, particularly in the context of renewable energy. The study, thereby, directly relates to the goals of eco-effectiveness and eco-efficiency postulated in Watson et al. (2010) as central objectives of Energy Informatics research. Therefore, Energy Informatics and the related concept of *Green IS* provide the theoretical anchor for our research within the IS community. The first part of this section is subsequently dedicated to work that relates to our approach in these fields of IS research. The second part addresses research on microgrids outside the IS discipline and elaborates on DER-CAM, the optimization module of our IS artifact.

2.1 Green IS and Energy Informatics

Green IS, as a subfield of Information Systems research, concerns the role of information systems in enabling, enhancing, and encouraging environmentally sustainable behavior and processes. While there has been pioneering research on these topics for several years, IS research on environmental sustainability gained particular traction within the community at the beginning of the current decade. Several publication in high-profile outlets emphasized the importance of Green

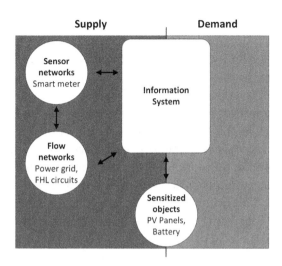

Figure III–1: The microgrid setting in the Energy Informatics Framework, based on Watson et al. (2010)

IS research, providing theoretical groundwork, and proposing research agendas (e.g. Melville, 2010; Watson et al., 2010; Elliot, 2011). Since then, various journal publications have further helped to position the discipline (e.g. vom Brocke et al., 2013; Malhotra et al., 2013) and to present promising Green IS applications (Seidel et al., 2013; Loock et al., 2013). This is complemented by a vast amount of conference publications, with Green IS research regularly featuring at all major IS conferences.

Our research project is an application of Energy Informatics research, as outlined in Watson et al. (2010). In their Energy Informatics Framework, they position the information system at the confluence of *flow networks* (circuits, power grids, pipelines, etc.), *sensor networks* (that report data on the flow network and environmental factors, such as a smart meter), and *sensitized objects* (that report data on their use and may be remotely controllable). Figure III–1 illustrates how the setting of the microgrid project reflects this framework. A difference to the original framework, as in Watson et al. (2010), is that sensitized objects in a microgrid are not limited to the demand side, since the PV panels can be remotely disconnected and the very purpose of the battery is the flexibility between supply and demand. The framework puts this setting in the context of eco-goals, which we have previously discussed, and the stakeholders. The interests of the stakeholders are particularly important in this project since they define both objective and constraints of the system. The U.S. government as operator of the base naturally seeks to reduce necessary operational spending. However, the public perception of "green" public buildings is similarly important to an administration that puts an increasing emphasis on sustainability. The utility company defines the energy tariff on the one hand and constrains the maximum energy that can be exported by the base on the other.

Naturally, there has been some research directly related to our approach within the Energy Informatics community in recent years. For instance, Feuerriegel et al. (2012) and Bodenbenner et al. (2013) analyze how IS design can contribute to the realization of demand response systems. These systems allow demand to be curtailed at certain times (*peak-clipping*) or moved to different times (*load-shifting*), thereby aligning it with energy supply. Brandt et al. (2013) introduce an information system that analyzes driving behavior to determine charging and discharging times for the battery of an electric vehicle. While our microgrid does not include electric vehicles, the battery installed at the base can provide additional supply or demand of energy on a flexible basis. These studies exemplify the contribution IS design can make to smart energy systems by providing coordination in the face of uncertainty and under organizational constraints.

2.2 General Research on Microgrids

Research on microgrids has been conducted for more than a decade (e.g. R. H. Lasseter & Paigi, 2004; Venkataramanan & Illindala, 2002). They are generally considered to be a solution to the problems and opportunities arising from the increasing decentralization of the power system due to renewable energy sources and distributed generation (R. Lasseter et al., 2002). During this time, two main research streams concerning microgrids have emerged. The first investigates control and coordination mechanisms for microgrids. For instance, Dimeas & Hatziargyriou (2005) present a multi-agent system for microgrid control, while Katiraei & Iravani (2006) analyze reactive strategies for power management. Publications in this context are either on a theoretical or conceptual basis. The second research stream investigates actual microgrid implementations. Examples include Nandi & Ghosh (2009), who evaluate a wind and battery based microgrid in Bangladesh, and Marnay et al. (2011), who discuss the microgrid at the Santa Rita Jail, a "green prison" in California. These implementations usually use very simple control strategies for the microgrid, which do not exploit its full potential.

This divergence between highly sophisticated control mechanisms, which have been developed on a theoretical level, and the simple management strategies that are actually implemented is due to the need for information of a certain quality to realize the former. For instance, the optimization of microgrid operations over a span of 48 hours requires reasonably reliable information on demand and energy generation within the microgrid. The design of such an information system that enables microgrid optimization under uncertainty is the research gap we address in this paper. As the actual optimization module within the information system, we employ DER-CAM (*Distributed Energy Resources Customer Adoption Model*). DER-CAM was originally designed as a tool to determine the optimal investment decisions for distributed energy resources, such as solar power, combined-heat-and-power plants, or batteries, in microgrids as well as for individual buildings. It is based on the *General Algebraic Modeling System* (GAMS). A new version of DER-CAM is also able to optimize microgrid operation over a certain time span (between 24 hours and one week), given information on future demand, generation, and prices. In this paper, we develop an information system that provides forecasts of these factors. It furthermore handles the resulting decisions of DER-CAM at each time step, given forecast uncertainty, as well as physical and organizational constraints. To our knowledge, this is the first time that microgrid implementations are considered from an IS perspective and that an actual microgrid implementation is investigated with respect to operational optimization under uncertainty.

3 Problem Statement and Research Design

This study is part of a project that establishes a microgrid at a U.S. army base. The occupancy of the base changes periodically between the permanent staff of about 250 residents and several thousand due to training schedules. The electrical network of the base is outlined in Figure III–2. Originally equipped with a one megawatt-peak (MWp) photovoltaic installation, another megawatt was added in late 2013, with a third currently being installed. Since the goal is to have a zero net-energy base, the solar installation will eventually reach 8 MWp. To stabilize grid operations, the utility company limits the possible export of PV power through the coupling point to one megawatt. If this amount is exceeded, parts of the PV installation would need to be disconnected, which can only be achieved in 0.5 MWp-segments. This is inefficient both from an ecological and economic perspective. Curtailing PV generation decreases the amount of clean

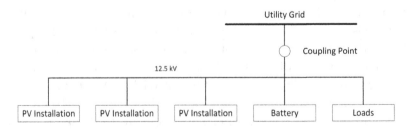

Figure III–2: Electrical network at the test site

Figure III–3: Research design of the microgrid project

energy produced by these installations. This energy also comes at virtually zero marginal cost, which enforces the economic argument. Since the power demand of the base usually varies between 0.8 and 1.8 MW, the threshold is regularly exceeded even by a 3 MWp installation and poses a major problem for any increase in installed power. Hence, a 1 MWh lithium-ion battery was installed to serve as a buffer for excess energy.

However, this battery can also be used for general demand shifts. The energy costs of the base are calculated according to the PG & E E-20 tariff structure (Pacific Gas &Electric Company, 2010), which designates different time-of-use rates for peak, part-peak and off-peak periods. More importantly, it also includes a demand charge for the highest peak, off-peak, and overall demand within a month, respectively. These demand charges account for a substantial share of total energy costs. Thus, it becomes evident that the management of the battery cannot be reduced to simple decisions, such as charge if there is excess PV generation or discharge if loads exceed PV generation. Instead, an accurate anticipation of future demand and generation, as well as an optimization according to these forecasts, are required to reduce the demand charges. The implementation of an information system that achieves all of these objectives is the general goal of the microgrid project.

The scope of this study within the project is illustrated in Figure III–3 and includes the design and evaluation of a prototype of the information system to determine its expected impacts before it is eventually implemented. Since our research is design-oriented, we follow the guidelines for design science research as outlined by Hevner et al. (2004). The *relevance of our problem* has been established in the preceding sections. We develop a tool that improves the integration of renewable energy sources into our power systems and increases the stability of the overall system. We *design as an artifact*—the IT artifact produced by our research is the resulting software package that addresses the aforementioned goals. Our *research contribution* is, therefore, the development of this novel application of IS design to support environmental sustainability, which will be *evaluated* later in this paper.

Our actual design process relates to a software engineering approach. First, we analyze the requirements our information system needs to fulfill. Second, we design the information system. Third, we evaluate the components of the IS design, and fourth, we evaluate the overall system and test if the initial requirements are satisfied. Given the intended objective of our information system, we derive the following requirements. The first three requirements are deduced from the agenda of the U.S. government as primary stakeholder, which have been discussed in the previous section. The fourth requirement is derived from the physical setting of the project and from the agenda of the utility company.

Requirement 1: Ecological efficiency. The primary goal of the project is to promote zero-net-energy consumption in government buildings. Hence, the information system must substantially reduce incidents when PV panels need to be shut off due to excess generation.

Requirement 2: Economic efficiency. While ecological efficiency is the primary objective, the information system also needs to reduce operational costs of the microgrid compared to simpler battery management strategies.

Requirement 3: Forecast accuracy. Since optimization occurs over a future timespan, the information systems must include components that supply the optimization module with forecasts of future demand and generation. Such forecasts must be reliable enough, so that optimization under uncertainty outperforms simpler strategies for battery management.

Requirement 4: Observance of constraints. The information system must account for organizational and physical constraints. These include contractual limits on exportable power, the fact that PV installations can only be disconnected discretely (500 kW segments), and the limits to the charging behavior of the battery.

In the following section we will present and evaluate the information system designed with these requirements in mind.

4 Information System Design and Evaluation

The information system we developed for the microgrid is illustrated in Figure III–4. The IT artifact at its core contains five components: two forecasting modules, one for the load of the base and one for the photovoltaic generation; the optimization module running DER-CAM; a database for historical data on generation, weather, and load; as well as a supervising module. The purposes of the latter are the coordination of the other modules and the communication with the components outside the artifact. The supervising module is written in Python and provides a link between the SQL-database, the GAMS-based optimization module, and the forecasting modules, which are currently running in R (although they will be moved to Python for the actual implementation, as well). The inputs from the smart meters and the weather forecast are collected by an energy management software (EMS) that also controls the battery and can (dis)connect PV installations. Our IT artifact receives the metering and weather data and enacts the schedules determined by the optimization module through an interface with this software.

The implementation of the information system results in the following processes. At fixed intervals, the EMS triggers the supervising module through the interface for an updated schedule. The supervising module requests the current weather forecast and metering data. These are collected by the EMS and forwarded to the supervising module. The

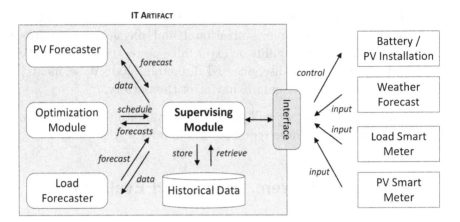

Figure III–4: Schematic representation of the information system

supervising module stores this new data in the database and retrieves and forwards historical datasets as required by the forecasters. The forecasting modules subsequently return the load and generation forecasts for the following 48 hours. The supervising module submits these values to the optimization module, which returns the optimal schedule given the forecasts. This schedule is eventually returned to and implemented by the EMS.

In the remainder of this section, we will first explain the methods behind the forecasting modules and the optimization module. Afterwards, we describe the setting used for the evaluation of the IT artifact. This is followed by the results, which include an evaluation of the forecasting components and of the overall system under different conditions.

4.1 Forecasting Modules

The **load forecaster** is based on the Discrete Fourier Transformation since the load curve shows very reliable patterns on a day-of-week basis. Building upon work by Hedwig et al. (2010), who use a Fast

Fourier Transformation (FFT) to forecast Wikipedia workload, which exhibits similarly stable patterns, we employ a FFT-algorithm (from the stats-package in R) to forecast the energy load for the army base. This is illustrated in Figures III–5a and III–5b. Figure III–5a shows the demand curves for three successive Thursdays (dark) and the dominant frequencies that have been extracted by the FFT-algorithm (light). Figure III–5b visualizes the fit between forecast (light) and actual load (dark) for the following Thursday.

While the fit is sufficiently good to serve as input for the optimizer, we mentioned the issue that the occupancy of the base sometimes changes substantially for several weeks at a time. While it would theoretically be possible to have an employee at the base provide information on the occupancy to the system, this presents several organizational challenges. On the one hand, it requires work hours that cannot be spent on other issues. On the other hand, occupancy of the base does not consistently translate to a certain increase or decrease of load, since it fundamentally depends on the activities of the occupants (e.g. computer-related vs. outdoor training). There is also the issue that a more detailed description of the occupants and their activities might touch sensitive information that should not be revealed to a microgrid operations information system. To overcome this problem, we introduce a learning parameter into the workload forecaster. Whenever deviations between forecast and actual load exceed a certain threshold for several successive periods, the part of the training dataset before these deviations started is scaled by a fraction of the average deviation in percent. This decreases the immediate forecasting error if base occupancy has changed, while limiting the impact of random deviations that are not related to an occupancy change (false positives).

The central assumption of our forecasting module for the PV generation is that power generated at time t, P_t^{PV}, on a clear day is directly related to the altitude of the sun above the horizon at that time, β_t (for $\beta_t > 0$). In the short term, this assumption is reasonable to provide a decent approximation, given the standard model by Masters

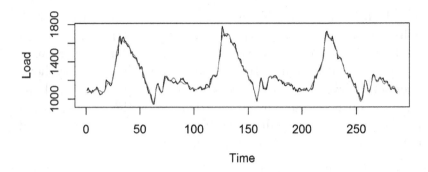

(a) Frequency extraction from historic dataset

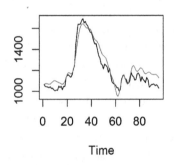

(b) Fit

Figure III–5: FFT-based load forecaster with time in 15-minutes intervals and load in kW

(2004). He calculates the radiation on a PV collector that tracks the movement of the sun as

$$P_t^{PV} \sim I_t$$
$$= A_t e^{-\frac{k_t}{\sin \beta_t}} \left[1 + C_t \left(\frac{1 + \sin \beta_t}{2} \right) + \rho_t (\sin \beta_t + C_t) \left(\frac{1 + \sin \beta_t}{2} \right) \right]$$

(III–1)

with ρ as the ground reflectance and A, k and C as the apparent extraterrestrial flux, the optical depth, and the sky diffuse factor, respectively. The latter three are seasonal variables that change slightly over the course of the year, but are basically constant in the short run. In fact, they change by substantially less than one percent between two successive days. However, the empirical data available for them is averaged over the continental United States and each month. Hence, the applicability of the values to our problem is doubtful and it is infeasible to collect them manually.

To solve this problem, we transform and simplify the model such that it can be estimated by a linear regression—including the seasonal variables, which are assumed to be constant factors in the short term. The terms in the brackets of Equation III–1 represent the multipliers for the direct, diffuse, and reflected radiation, respectively. Direct radiation usually accounts for 80 to 90 percent of the total (Masters, 2004), hence we ignore the diffuse and reflected parts for approximation, which reduces the model for short-term considerations to

$$P_t^{PV} \sim A e^{-\frac{k}{\sin \beta_t}}.$$

(III–2)

Taking the logarithm of both sides yields

$$\ln P_t^{PV} \sim \ln A - k \frac{1}{\sin \beta_t}.$$

(III–3)

Hence, the relationship can be estimated through the linear regression

$$y_t = \ln P_t^{PV} = a_0 + a_1 \frac{1}{\sin \beta_t}.$$

(III–4)

Since our assumptions only hold in the short term, the training sample for our regression contains data of the thirty most recent clear-sky daylight ($\beta_t > 0.05$) hours. If the interval to be forecasted contains points with cloudy or hazy weather, the clear-sky prediction is multiplied by a certain factor, depending on whether the forecasted weather is overcast (including rain / snow), partially clouded, or foggy. Since a qualitative phrase like "overcast" covers a large spectrum of conditions, these factors are adjusted dynamically if deviations, i.e. $[\mathbb{E}(P_t^{PV}) - P_t^{PV}]/P_t^{PV}$, exceed a certain threshold.

One issue that could potentially affect the precision of this forecasting model is the fact that it forecasts the logarithm of expected PV generation, which would need to be inverted to receive the expected PV generation the optimizer can work with. Since the linear regression weights a deviation between $y_t = 0.2$ and $\mathbb{E}(y_t) = 0.3$ the same as the deviation between $y_t = 6.0$ and $\mathbb{E}(y_t) = 6.1$, the model could result in substantial estimation errors in the upper ranges. To solve this, we approximate a second model that relates P_t^{PV} linearly to $\sin \beta_t$ (the derivation of this model is attached in Section 8). While the first model better describes the functional form, the approximation in the second model reduces the impact of forecasting errors.

PV Model 1: $\mathbb{E}(\ln P_t^{PV}) = a_0 + a_1 \frac{1}{\sin \beta_t}$

PV Model 2: $\mathbb{E}(P_t^{PV}) = a_0 + a_1 \sin \beta_t$

Both models will be assessed during the evaluation later in this section.

4.2 Optimization Module

The optimization module uses the CPLEX-solver in GAMS to solve the following minimization problem. The variables and parameters are described in Table III–1.

Equation III–5 states the loss-minimizing objective function. The loss is defined as the sum of all demand charges and the energy costs over two days, measured in 15-minute intervals (for a total of 192 periods).

Table III–1: Variables and parameters in the optimization problem

Symbol	Unit	Explanation
$\mathbf{b} =$ $(b_1 \quad b_2 \quad \ldots \quad b_{192})$	kWh	Vector for the change in the amount of energy stored in the battery for each period t.
$\mathbf{z} =$ $(z_1 \quad z_2 \quad \ldots \quad z_{192})$	None	Vector for the Photovoltaic segments that are disconnected in each period t.
E_t	kWh	Energy balance with utility grid in period t (negative numbers imply that energy was sold to the utility).
p_t	USD/kWh	Price of energy in period t (defined through contract with utility).
$M^{peak}, M^{part}, M^{total}$	kW	Maximum demand during peak hours, part-peak hours, or any hours, respectively.
c_1, c_2, c_3	USD/kW	Demand charge for maximum demand during peak hours, part-peak hours, or any hours, respectively.
L_t, PV_t	kWh	Load and photovoltaic generation in period t.
B_t	kWh	Energy stored in the battery at the beginning of period t.
B^{min}, B^{max}	kWh	Maximum and minimum amounts of energy that can be stored in the battery, respectively.
b^{dc}, b^c	kW	Maximum discharging and charging power of battery.
ω	None	Charging and discharging efficiency (assumed to be equal).
PV_t^{eff}	kW/kWp	Photovoltaic efficiency in period t, i.e. power generated per power installed.
PV^{inst}	kWp	Installed photovoltaic power
s_t, r_t	None	Booleans, equal to 1 if t is a peak period or part-peak period, respectively.

The tariff follows net metering, i.e. energy is sold to the utility at the same price that it is purchased from the utility.

$$L = \min_{b,z} \left[c_1 \cdot M^{peak} + c_2 \cdot M^{part} + c_3 \cdot M^{total} + \sum_{t=1}^{192} p_t \cdot E_t \right] \quad \text{(III–5)}$$

The energy balance with the utility is the difference between load and photovoltaic generation plus any change in the amount of energy stored in the battery. Depending on whether the change is negative or positive, it has to be multiplied by the charging efficiency or its inverse, respectively (Equations III–6 and III–7).

$$\text{s.t.} \quad E_t = L_t - PV_t + \gamma_t \cdot b_t \quad \forall t \quad \text{(III–6)}$$

$$\gamma_t = \begin{cases} 1/\omega & \text{if } b_t \geq 0 \\ \omega & \text{otherwise} \end{cases} \quad \forall t \quad \text{(III–7)}$$

The amount of energy stored in the battery in $t+1$ is the amount stored in t plus the change b_t (Equation III–8). B_1, i.e. the energy at the beginning of the optimization, is provided by the supervising module.

$$B_{t+1} = B_t + b_t \quad \forall t \quad \text{(III–8)}$$

Equations III–9 and III–10 define limits to the amount of energy that can be stored in the battery and to the change of that amount within a single period.

$$B^{min} \leq B_{t+1} \leq B^{max} \quad \forall t \quad \text{(III–9)}$$

$$b_t \in \{ \mathbb{R} \mid b^{dc} \cdot 0.25 \leq b_t \leq b^c \cdot 0.25 \cdot \omega \} \quad \forall t \quad \text{(III–10)}$$

The maximum power export is 1000 kW, equal to 250 kWh in a 15-minute interval.

$$E_t \geq -250 \quad \forall t \quad \text{(III–11)}$$

Equation III–12 states that photovoltaic generation in t is the efficiency times the installed capacity. The latter may be reduced by z_t segments (500 kW each) to meet Constraint III–11.

$$PV_t = PV_t^{eff} \cdot \left(PV^{inst} - 0.5z_t\right) \cdot 0.25 \qquad \forall t \qquad \text{(III–12)}$$

$$z_t \in \{\mathbb{Z} \quad | \quad 0 \le z_t \le 2PV^{inst}\} \qquad \forall t \qquad \text{(III–13)}$$

Peak loads are determined as below. The initial values M_0^{peak}, M_0^{part}, M_0^{total} are provided by the supervising module and the respective demand maxima during periods before $t = 1$ that are in the same billing month.

$$M^{peak} = \max\left(0, M_0^{peak}, (E_1 \cdot 4 \cdot s_1), (E_2 \cdot 4 \cdot s_2), \dots, (E_{192} \cdot 4 \cdot s_{192})\right)$$
$$\text{(III–14)}$$

$$M^{part} = \max\left(0, M_0^{part}, (E_1 \cdot 4 \cdot r_1), (E_2 \cdot 4 \cdot r_2), \dots, (E_{192} \cdot 4 \cdot r_{192})\right)$$
$$\text{(III–15)}$$

$$M^{total} = \max\left(0, M_0^{total}, (E_1 \cdot 4), (E_2 \cdot 4), \dots, (E_{192} \cdot 4)\right) \qquad \text{(III–16)}$$

The schedule produced by the optimization module is implemented as is, which likely results in some sub-optimal decisions due to forecast errors. The only exception is that the battery is not charged if that would increase the current maximum demand, since the likelihood of such a decision being based on erroneous forecasts is very high. In the following evaluation, we will analyze whether the optimization module outperforms simpler decision strategies despite those forecast errors.

4.3 Evaluation Setting

While the optimization module requires data in 15-minutes intervals, the historical data on photovoltaic generation at the base only contains hourly observations. The exception to this is a two week long test study in January 2013, which includes data in the required 15-minutes intervals. These observations serve well to test the components and the overall system presented in this paper. At the time, only one megawatt of photovoltaic capacity was installed. However, we can easily scale this amount to reflect current and future situations since it can be reasonably assumed that the generated power increases approximately linearly in the installed capacity. We split the datasets into two subsets. The first week is the training set used to estimate clear-sky generation for the first day of the second week, the test set. Once these days have passed, the observations are added to the training set to improve the forecast for future days.

Historical information on the cloud cover on those particular days is only available for every full hour and in qualitative form—categories, such as "clear", "light rain", "overcast", or "haze". We linked this qualitative information to the half hour before and after the respective full hour. For instance, if the reported condition is "clear" at 3 p.m. and "light rain" at 4 p.m., the intervals 2:30–2:45, 2:45–3:00, 3:00–3:15, and 3:15–3:30 would be labeled "clear", with the following four intervals until 4:15–4:30 being labeled "light rain". Naturally, this introduces a lot of uncertainty into the data, since it is unclear whether the reported condition prevailed for the entire hour. Also, qualitative data is by definition fuzzy. A label such as "overcast" can include a wide spectrum of situations, from a cloudy sky with occasional sunny spots to a sky filled with dark-gray clouds that suppress virtually all photovoltaic generation. However, the information system will eventually work with weather forecasts, which have their own inherent uncertainty. Nevertheless, it is likely that they are going to provide a more accurate prediction than the historical data we are working with for prototype testing due to two reasons. First, weather forecasts are quite accurate for 24-hour predictions and even 48 hours (although becoming more unreliable after that). Second, weather forecasts include information on the actual cloud cover in percent, which provides much more accuracy than the qualitative information of the historical dataset. Hence, the results of our evaluation should be considered as lower bounds of the actual impact of the information system once it is implemented.

4.4 Evaluation Results

We will first evaluate the forecasting modules of the information system for the test week from 2013-01-25 to 2013-01-31. Afterwards, we will analyze the entire system, once for 3 megawatt-peak installed photovoltaic power (the situation when the system will be implemented) and once for 5 megawatt-peak.

Evaluation of the Load Forecaster

Table III–2: Comparison of forecasted and actual loads (all values in percent)

	Average relative deviation	*Share with deviation < 10%*	*Deviation over entire period*
Day 1	7.71	73.96	2.84
Day 2	8.23	83.33	2.89
Day 3	5.79	80.21	3.92
Day 4	6.36	86.46	1.34
Day 5	4.30	90.63	0.26
Day 6	3.48	95.83	0.53
Day 7	9.22	61.46	7.46
Week	6.44	81.70	0.61

Since the load data is available in 15 minute intervals for several months, as opposed to the PV data, we could implement the load forecaster as explained earlier in this section. We forecasted the load for each day using a Fast Fourier Transformation of the load data for the three preceding days that share the same day-of-week. Holidays were counted as Sundays, regardless of their actual day-of-week (New Year's Day and Martin Luther King Jr. Day fell in our training set).

Table III–2 suggests that the FFT-based forecasting works very well and confirms the visual comparison for a single day in Figure III–5. The first column lists the average relative deviation over all 96 observations for each day (with the last row for the entire week). The deviation is calculated as $\left|(\hat{L}_t - L_t)/L_t\right|$, with $\hat{L}_t$ and L_t as the forecasted and actual load values, respectively. The average deviation is consistently below ten percent for most days and substantially so for the week as a whole. This is enforced by the second column, which shows the share of observations with a deviation less than ten percent. With the exception of Day 7, this share is above 70%, indicating that the vast majority of estimates are almost spot on. When we consider the deviation over the entire period, i.e. between the forecast and actual values for the entire day / week, we can observe that these are very low, as well. It is especially noteworthy that the deviation over the entire week is almost zero. In summary, the load forecaster should serve our purpose quite well.

Evaluation of the PV Forecaster

Earlier in this section, we derived two regression models to forecast clear-sky photovoltaic generation. The first relates the natural logarithm of generation to the reciprocal of the sine of the sun altitude angle. While we do focus on clear-sky generation, the uncertainty introduced by the historical data on cloud cover still prevails, as illustrated in Figure III–6. The light gray columns from 9:30 a.m. on January 29 to midnight on January 30 indicate that "clear sky" was reported for the entire interval. However, the generation curve for the first day is much more volatile, suggesting that clouds were still present that were not captured by the qualitative data. Since the forecast errors thus introduced into the regression model may be amplified in our logarithmic model, we derive a second model in Section 8 that relates photovoltaic generation directly to the sine of the sun altitude.

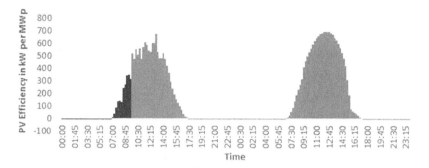

Figure III–6: Actual PV efficiency for Jan 29 and Jan 30, 2013. Light gray columns indicate reported clear sky.

Similar to the load forecaster, we calculate the deviations between predicted and actual clear-sky generation to evaluate the models. The results are summarized in Table III–3. Since photovoltaic generation exhibits a much higher coefficient of variation than load, we also consider the absolute deviations, in addition to the relative deviations. After all, for the optimization module a forecasted value of 500 for an actual generation of 250 is much worse than a forecasted value of 40 given an actual generation of 20—although the relative deviation is the same in both cases. Table III–3 shows that the second, linear model substantially outperforms the logarithmic model. On average, the absolute deviations in the linear model are about a third as high compared to the logarithmic model. This result is amplified when considering mean squared errors, where the linear model outperforms the logarithmic model by a factor of eight. While the relative deviation in the linear model is much higher than for the load forecaster, this was to be expected due to the uncertainty introduced by the qualitative data on cloud cover. Also, the fourth row shows that much of this uncertainty is caused by forecasts of low actual generation. Since the battery is likely needed during times of high generation—when the

Table III–3: Comparison of forecasted and actual PV generation (per MWp installed) for each model

	Model 1 (logarithmic)	Model 2 (linear)
Avg. absolute deviation (kW)	176.01	61.13
Avg. relative deviation (percent)	42.17	16.71
Mean Squared Error	57,701	7,400
Avg. relative deviation when actual PV > 600 kW (percent)	37.58	5.71

excess power exceeds the export limit—the second model produces very reliable results given the inherent uncertainty of the data for these times.

In summary, the forecast of photovoltaic generation is more difficult and less reliable than the load forecast. Nevertheless, given the uncertain qualitative data on cloud cover, the linear model provides quite accurate predictions, particularly for times of peak production, which are the most relevant to battery management. Hence, we will employ the linear model in the evaluation of the entire system.

System Evaluation for 3 MWp Installed

As there will be a photovoltaic installation with three megawatt peak power at the army base when the information system is implemented, we use this as the first case of our system evaluation. Recall the expectations for the information system we derived in the requirements analysis. We established that *forecast accuracy* is given, with both forecasters providing quite accurate predictions. The quality of the photovoltaic forecast should be a lower bound, since quantitative day-ahead weather forecasts are likely to be more precise than the qualitative historical data we work with. The *constraints*, such as battery and power export limitations, have been included in the optimization problem. This leaves *economic* and *ecological efficiency* as the requirements to be evaluated in this subsection. We compare the performance of our IT artifact to two benchmark cases. The first benchmark is a scenario without a battery as an energy buffer. The information system manages the battery charging strategies, so it is natural to compare it to this situation and assess the ecological effect of an IS-managed battery system. However, Brandt et al. (2013) have suggested that simple decision strategies for batteries may outperform optimization models when facing uncertainty. Their case study only considers a single household and electric vehicle, resulting in a much more volatile load and battery availability than in our case. Nevertheless, we choose such a simple decision strategy as our second benchmark to evaluate economic efficiency. According to this strategy, the battery always charges if excess PV generation exceeds the exportable amount, unless the battery is full. The battery is discharged if the load of the base exceeds PV generation.

Table III–4 summarizes the evaluation results and provides several interesting insights. First, for the three megawatt-peak installation in January, the battery is barely necessary if the focus is on ecological efficiency. During the course of the week, only six PV segments need to be turned off because the export limit is exceeded. Naturally, the benefit of the simple decision strategy is very limited. The optimized schedule provided by our IT artifact on the other hand uses the battery to decrease demand charges even when the PV generation does not exceed the export limit. The reduction of those charges is quite substantial, with a drop of more than USD 1,100. Since they are calculated on a monthly level and we only analyze one week, we weighted them accordingly to arrive at an overall drop of 2.32% in total energy costs. These first results confirm that a simple decision strategy is not suitable for the microgrid since demand charges are only reduced if the times of highest demand and battery discharging randomly coincide. Since demand charges are a substantial part of

Table III–4: Results of system evaluation for 3 MWp installed

	Optimization	Decision Strategy	No Battery
Energy costs (USD)	10,031.89	10,028.29	10,070.32
Demand charge (USD)	13,850.73	14,977.48	14,977.48
Weighted sum (USD)	13494.57	13772.66	13814.69
Weighted sum (percent)	97.68	99.70	100.00
Disconnected PV segments	0	0	6

total energy costs—even more so when PV output increases—the optimization approach is generally superior.

In the next step, we increase the installed power to five megawatt-peak. Thereby, we can get a better grasp of the ecological benefits when more panels are installed or during the summer months, when generation is generally higher.

System Evaluation for 5 MWp Installed

When we consider an increase of installed PV power to five megawatt-peak—in the medium run even an increase to eight is planned—the ecological benefits of the IS-controlled battery system become more evident. The results in Table III–5 illustrate that the number of PV segments that need to be disconnected is reduced by 78, or 22 percent. Over the course of the week, the additional energy that could be generated this way added up to six megawatt-hours or an increase of 5.4% compared to the situation without batteries.

The maximum demand charge is still substantially reduced. However, the information system now also uses excess photovoltaic generation

Table III–5: Results of system evaluation for 5 MWp installed

	Optimization	*No Battery*
Energy costs (USD)	7,344.23	7,799.55
Demand charge (USD)	13,769.61	14,731.52
Weighted sum (USD)	10,786.64	11,482.43
Weighted sum (percent)	93.94	100.00
Disconnected PV segments	280	358

to drive down energy costs. Overall, weekly energy costs are reduced by USD 700 or six percent.

5 Discussion

In this section, we first discuss the technical and economic implications of our work. This is followed by suggestions on the future role of Information Systems research with regard to Microgrids and Smart Power Systems.

5.1 Technical and Economics Implications

The evaluation in the previous section illustrates that the information system we designed contributes both to the economic and ecological efficiency of the microgrid. Despite the inherent uncertainty of forecasted load and PV generation, the information system allowed for an additional six megawatt-hours of photovoltaic energy to be generated over the course of the week. Consider that the minimum charge of the battery is 20 percent, charging efficiency is 0.92, and on a typical day the battery successively charges during the morning and afternoon, and discharges in the evening. The resulting theoretical maximum of additional PV energy over the week is around 6.6 megawatt-hours. Hence, the ecological efficiency of our system is very close to the theoretical maximum, especially when considering that the first day in our sample was very cloudy.

While not the primary focus of the project, it is interesting to consider the economic implications of our results. Our information system uses the battery not only to store excess photovoltaic power, such as a comparable trigger strategy would do as shown above, but also to actively keep demand charges down. We only analyze a sample week in January, but we can use these results as a lower bound for the entire year. Cost savings are likely higher during the summer since prices for energy and demand are higher and the number of sunny days increases.

As Stadler et al. (2013) point out, stationary battery technology is still too expensive for the costs to be recovered. Consequently, we first analyze how much our information system contributes to cost recovery given current technologies. While the costs for the actual battery at the base cannot be disclosed, PG&E recently launched a battery storage system with four megawatts of power at USD 3.3 million (Kligman, 2013). Hence, the current costs of a system similar to the one we considered (1.25 MW) should be around USD 1 million. The lifespan of the PG&E system is about 15 years. If we consider the weekly savings of the 5 MWp scenario at USD 700 to be representative of the year (as discussed above, it is more likely to be a lower bound) and a discount rate of five percent, about 41% of investment costs are recovered during the lifetime of the battery.

However, utility-scale battery technology is rapidly evolving. The company EOS Energy Storage intends to launch a zinc-air based battery by 2015 at a price of USD 200 to USD 250 per kWh (Parkinson, 2013) and a lifespan of up to 30 years. Even assuming battery costs of USD 300 per kWh and annual operational costs of USD 10,000 (e.g. for IS operation), investment costs would be recovered after 15 years—half the lifespan of the battery. As we considered a lower bound for energy savings, the actual break-even point may arrive even sooner.

Environmental sustainability in itself is an important goal for our society and in this paper we show from a technical perspective how information systems can contribute to that goal by increasing the efficiency of renewable energy systems. However, the examples discussed in this section also show that they increase the economic feasibility of these solutions. With the rapid technological advancement of utility-scale batteries, information systems, such as the one we designed in this paper, will be necessary to enable the full potential of microgrids. Thereby, they make the investment in sustainable energy technologies not only the ecological but also the economic thing to do.

5.2 Implications for IS Research

While this work focuses to a substantial degree on the technical aspects of implementing an information system in a Microgrid context, we must also acknowledge the possible contribution of IS research with respect to the socio-economic environment of a future power system. After all, this technical focus has only been feasible because the agendas of the stakeholders in our case study were quite straightforward and aligned with each other. As discussed earlier, the primary goal of the U.S. government as operator of the base is to increase the ecological efficiency of the photovoltaic system installed there. Due to the experimental and showcase nature of the project, economic considerations are not as binding as they may be in other situations. The utility company as the second stakeholder in the system seeks to reduce the variance of photovoltaic power fed into the grid by the base. However, this objective is neatly codified in the power export limit of one megawatt. The base itself follows strict schedules and clear hierarchies. Altogether, this study is very close to the ideal case to assess the technical benefits of an energy information system.

The necessity of future IS research is emphasized once this environment becomes more complicated and complex, as for microgrids in large office buildings or residential neighborhoods. The number and heterogeneity of stakeholders substantially increases and economic considerations become a decisive factor. For instance, in the case of a residential microgrid, households with photovoltaic panels and other forms of energy generation may have different objectives than households without these devices. The distribution of costs and benefits of the microgrid must provide sufficient incentives to participate to all agents. Once this social aspect is taken into consideration, there are several streams in IS research that may provide valuable contributions in the future.

Service Research. As illustrated by the power export limit in our case study, the agendas of the stakeholders must be codified in some contractual form. The benefit of a household to shift demand in a manner beneficial to the overall microgrid must be clearly stated. This gives rise to a variety of new energy services (Strueker & Dinther, 2012) which can build upon established research on IT services.

Privacy Research. In our case study, the operators of the microgrid, of the PV installation, and of all demand loads were the same entity. Once this changes, privacy issues become a valid concern, as agents may not be willing to share all information necessary for an optimal operation of the microgrid.

Human-Computer Interaction. The army base in our case study followed to a substantial extent strict, centrally organized schedules. On the contrary, a microgrid of residential households needs to take into account and affect the behavior of many heterogeneous agents. This requires more interaction between the smart microgrid and the users. IS research can provide insights on interfaces, designs, and strategies for this interaction, as exemplified by the Velix energy management system in Loock et al. (2013).

6 Conclusion

On the path towards an environmentally sustainable energy supply, microgrids are considered to be a key technology. They enable the integration of distributed renewable energy sources and increase the robustness of the overall power system. However, an efficient coordination of the microgrid requires information about future supply and demand, and needs to incorporate organizational, contractual, and physical constraints. Research on the corresponding information systems is mandatory to successfully overcome this challenge.

The goal of this paper was to develop and evaluate a prototype of such an information system, so that it can eventually be implemented as part of a pilot study in a U.S. military base. More specifically, our research questions related to the requirements the information systems needs to meet, the actual information system design, and whether the resulting IT artifact satisfies those requirements. Based on the project setting and the agendas of the project stakeholders, we determined that the IT artifact must improve the ecological efficiency of the system by reducing the number of incidents when parts of the photovoltaic installation need to be disconnected. Furthermore, it needs to be able to forecast uncertain information on load and generation sufficiently accurately and incorporate organizational and physical constraints. Finally, it needs to reduce the overall costs of the battery system compared to less sophisticated management strategies.

The IT artifact we designed consists of several modules, including two for forecasting (load and photovoltaic generation), one for optimization, and one supervising module. While the constraints were incorporated into the optimization problem, our evaluation showed that the other requirements are fulfilled, as well. Both forecasting modules exhibit low prediction errors, with the error of PV forecasts being slightly higher due to qualitative weather data. We could also show that our information system is able to directly target and reduce demand charges, making it economically superior to simple trigger strategies.

Most importantly, the primary goal of ecological efficiency is achieved. Over the course of our test week, the system enabled the generation of six megawatt-hours of additional clean photovoltaic energy. We could show that for a typical week with the battery charging during the day and discharging in the evening, this is very close to the theoretical maximum.

The evaluation also illustrated that with improving technology and falling battery prices, investments in energy storage may soon turn out to be not just the best decision from an ecological, but also an economic perspective. We estimated that even in a worst-case scenario, battery prices of USD 300 per kWh can be recovered well before the end of the lifespan of the battery, including operational costs. Hence, the implications of our results go beyond this pilot study. We could demonstrate the improvements in ecological and economic efficiency large building complexes can achieve through IS-supported microgrids. Given the anticipated reductions in prices of energy storage, this provides a viable model for government facilities, but also large office buildings.

Naturally, the next step for our research is to implement the information system at the base. Given the positive assessment of the prototype despite very conservative assumptions, we will evaluate the system to determine its ecological and financial potential over the course of the year. This should provide an even better understanding of how IS research can aid public institutions in reducing their environmental footprint and in promoting sustainability.

7 References

Bodenbenner, P., Feuerriegel, S. & Neumann, D. (2013). Design Science in Practice: Designing an Electricity Demand Response System. In D. Hutchison et al. (Eds.), *Design Science at the*

Intersection of Physical and Virtual Design (Vol. 7939, pp. 293–307). Berlin, Heidelberg: Springer Berlin Heidelberg.

Brandt, T., Feuerriegel, S. & Neumann, D. (2013). Shaping a Sustainable Society: How Information Systems Utilize Hidden Synergies between Green Technologies. *ICIS 2013 Proceedings*, Paper 7.

Chaurey, A. & Kandpal, T. C. (2010). A techno-economic comparison of rural electrification based on solar home systems and PV microgrids. *Energy Policy, 38*(6), 3118–3129.

Dimeas, A. L. & Hatziargyriou, N. D. (2005). Operation of a Multiagent System for Microgrid Control. *IEEE Transactions on Power Systems, 20*(3), 1447–1455.

Elliot, S. (2011). Transdisciplinary Perspectives on Environmental Sustainability: A Resource Base and Framework for IT-Enabled Business Transformation. *Management Information Systems Quarterly, 35*(1), 197–236.

Feuerriegel, S., Strüker, J. & Neumann, D. (2012). Reducing Price Uncertainty through Demand Side Management. *ICIS 2012 Proceedings*, Paper 7.

Goebel, C., Jacobsen, H.-A., Razo, V. d., Doblander, C., Rivera, J., Ilg, J., ... Lässig, J. (2014). Energy Informatics - Current and Future Research Directions. *Business & Information Systems Engineering, 6*(1), 25–31.

Hedwig, M., Malkowski, S. & Neumann, D. (2010). Towards Autonomic Cost-Aware Allocation of Cloud Resources. *ICIS 2010 Proceedings*, Paper 180.

Hevner, A., March, S., Park, J. & Ram, S. (2004). Design Science in Information Systems Research. *Management Information Systems Quarterly, 28*(1), 75–106.

Katiraei, F. & Iravani, M. R. (2006). Power Management Strategies for a Microgrid With Multiple Distributed Generation Units. *IEEE Transactions on Power Systems, 21*(4), 1821–1831.

Kligman, D. (2013). *Largest Battery Energy Storage System in California to Improve Electric Reliability for Customers.* Retrieved on April 13, 2015, from http://www.pgecurrents.com/2013/05/23/largest-battery -energy-storage-system-in-california-to-improve -electric-reliability-for-customers/

Kwasinski, A., Krishnamurthy, V., Song, J. & Sharma, R. (2012). Availability Evaluation of Micro-Grids for Resistant Power Supply During Natural Disasters. *IEEE Transactions on Smart Grid, 3*(4), 2007–2018.

Lasseter, R., Akhil, A., Marnay, C., Stephens, J., Dagle, J., Guttromson, R., ... Eto, J. (2002). *Integration of Distributed Energy Resources: The CERTS MicroGrid Concept: Report prepared for the US Department of Energy.*

Lasseter, R. H. & Paigi, P. (2004). Microgrid: a conceptual solution. *PESC'04 Proceedings, 6*, 4285–4290.

Loock, C.-M., Staake, T. & Thiesse, F. (2013). Motivating Energy-Efficient Behavior with Green IS: An Investigation of Goal Setting and the Role of Defaults. *Management Information Systems Quarterly, 37*(4), 1313–1332.

Malhotra, A., Melville, N. & Watson, R. T. (2013). Spurring Impactful Research on Information Systems for Environmental Sustainability. *Management Information Systems Quarterly, 37*(4), 1265–1274.

Marnay, C., DeForest, N., Stadler, M., Donadee, J., Dierckxsens, C., Mendes, G., ... Cardoso, G. F. (2011). A Green Prison: Santa Rita Jail Creeps Towards Zero Net Energy (ZNE). *ECEEE 2011 Summer Study.*

Masters, G. M. (2004). *Renewable and Efficient Electric Power Systems.* Hoboken, NJ: Wiley & Sons, Inc.

Melville, N. (2010). Information Systems Innovation for Environmental Sustainability. *Management Information Systems Quarterly, 34*(1), 1–21.

Nandi, S. K. & Ghosh, H. R. (2009). A wind–PV-battery hybrid power system at Sitakunda in Bangladesh. *Energy Policy, 37*(9), 3659–3664.

Paatero, J. V. & Lund, P. D. (2007). Effects of large-scale photovoltaic power integration on electricity distribution networks. *Renewable Energy, 32*(2), 216–234.

Pacific Gas &Electric Company. (2010). *Electric Schedule E-20.* Retrieved on April 13, 2015, from `http://goo.gl/CIs8jB`

Parkinson, G. (2013). *EOS: utility scale battery storage competitive with gas.* Retrieved on April 13, 2015, from `http://reneweconomy.com.au/2013/eos-utility-scale -battery-storage-competitive-with-gas-36444`

Seidel, S., Recker, J. & Vom Brocke, J. (2013). Sensemaking and Sustainable Practicing: Functional Affordances of Information Systems in Green Transformations. *Management Information Systems Quarterly, 37*(4), 1275–1299.

Stadler, M., Kloess, M., Groissböck, M., Cardoso, G., Sharma, R., Bozchalui, M. C. & Marnay, C. (2013). Electric storage in California's commercial buildings. *Applied Energy, 104*, 711–722.

Strueker, J. & Dinther, C. (2012). Demand Response in Smart Grids: Research Opportunities for the IS Discipline. *AMCIS 2012 Proceedings*, Paper 7.

Venkataramanan, G. & Illindala, M. (2002). Microgrids and sensitive loads. *2002 IEEE Power Engineering Society Winter Meeting*, 315–322.

vom Brocke, J., Watson, R., Dwyer, C., Elliot, S. & Melville, N. (2013). Green Information Systems: Directives for the IS Discipline. , Article 30.

Watson, R., Boudreau, M.-C. & Chen, A. (2010). Information Systems and Environmentally Sustainable Development: Energy Informatics and New Directions for the IS Community. *Management Information Systems Quarterly, 34*(1), 23–38.

8 Appendix

Derivation of Model 2 for the PV forecasting module:

Recall Equation III–2:

$$P_t^{PV} \sim A e^{-\frac{k}{\sin \beta_t}}.$$

In the interval $[0.05, 0.975]$ (we set expected generation to zero for angles less than 0.05, since it is either night or early dawn / late dusk), we can approximate $\ln(\sin \beta_t)$ reasonable well as $\frac{-p}{\sin \beta_t} + q$, with p and q as some constants, particularly with $p \sim 0.18$. This yields

$$P_t^{PV} \sim A e^{\frac{k}{p} \ln \sin \beta_t + \frac{kq}{p}},$$

which in turn translates to

$$P_t^{PV} \sim A(\sin \beta_t)^{\frac{k}{p}} + z,$$

with $z = A e^{\frac{kq}{p}}$. As Masters (2004) points out, k varies between 0.142 and 0.207. With $p \sim 0.18$, the exponent is very close to 1 (given the short term relevance of our model, we can even set $p = k$ and still find a q that provides a good approximation), such that:

$$P_t^{PV} \sim z + A \sin \beta_t.$$

This relationship can be estimated through a linear model. A similar result can be reached through a Taylor Expansion of the right-hand side of Equation III–2.

IV A Business Model for Employing Electric Vehicles for Energy Storage

Working Paper. Parts of the paper have been published in:

Tobias Brandt, Sebastian Wagner, and Dirk Neumann, "Road to 2020: IS-Supported Business Models for Electric Mobility and Electrical Energy Markets" (2012), ICIS 2012 Proceedings, Paper 48.

Sebastian Wagner, Tobias Brandt, and Dirk Neumann, "Beyond Mobility - An Energy Informatics Business Model for Vehicles in the Electric Age" (2013), ECIS 2013 Proceedings, Paper 1.

Abstract

Aggregating the storage capabilities of electric vehicles is generally considered to be a promising method of supporting the integration of volatile renewable energy sources into the power grid. We analyze this concept from an economic perspective by developing and evaluating a business model for electric vehicle aggregation. Specifically, we investigate the case of parking garage operators using the electric vehicles located at their facilities to provide reserve energy for frequency regulation. We evaluate revenues and cost structures using extensive real-world data sets on the German market for frequency regulation, on battery states of charge at different times of day, and on occupancy rates of parking facilities. We find that possible revenues given current market conditions are far inferior to investment costs for charging and IT infrastructure. Even if the operator installs these features to enable customers to charge their vehicles for a fee, it would be more profitable to focus on this service and charge vehicles immediately when they enter the garage instead of delaying charging for frequency regulation. Overall, we conclude that a large-scale adoption of electric mobility may produce greater challenges to the power grid than is currently projected.

1 Introduction

During the past decade, electric mobility has received attention as a promising way to reduce transportation-related carbon emissions. However, actual sales figures are only slowly increasing, often depending on national subsidy schemes to bridge the price difference between electric and conventional vehicles. A proposed strategy to soften the financial strain on prospective customers is to use electric vehicles to store energy generated by intermittent renewable sources. The idea is intriguing since their dependence on exogenous factors, such as wind or sun, is a crucial drawback of many renewable energy sources. Furthermore, cars are generally only used during a small portion of the day, such that electric cars could be used to store energy and return it to the grid as needed during the remaining hours. The owners of these electric vehicles (EVs) would receive a monetary compensation for this service. Thus, the financial burden for prospective buyers is decreased. At the same time, the grid integration of renewable energy sources is improved, resulting in a win-win situation for vehicles owners and grid operators.

In this paper, we investigate the question, why the concept of using electric vehicles for energy storage (EV4ES) has not yet taken root in the real world, despite these optimistic prospects. As we will outline shortly, there is a substantial body of research on the EV4ES concept. While most of past research throughly analyzes the technical feasibility, economic analyses are often based on extensive assumptions. This reliance on assumptions results in a high variance of expected revenues, ranging from a few dollars to several thousand dollars per car and year. By contrast, we construct a comprehensive business model for such an EV4ES implementation to model costs and use simulations based on real-world data to assess the main revenue streams. Through the following three research questions, we investigate whether EV4ES is not only technologically feasible but also economically viable.

What constitutes a **reasonable business model** for EV4ES?

What are the **associated costs** and **expected revenue streams**?

What **managerial and policy implications** can be derived from the results?

The first question is arguably subjective since different people may perceive different business models as reasonable. However, as we will outline, various studies have used parking garage operators as intermediaries in EV4ES schemes. We will extensively draw on literature on business models to further investigate this proposal (e.g. Osterwalder, 2004; Alt & Zimmermann, 2014; Veit et al., 2014). For the second research question, we conduct a simulation analysis that utilizes research on the technical foundations of EV4ES on the one hand (Kempton & Tomić, 2005a) and real-world data on the other. Finally, we discuss our results to address the third research question.

This paper proceeds as follows. In Section 2, we outline the research framework and summarize work that relates to our approach. In Section 3, we model the business case in order to address the first research question. Section 4 contains the simulation of the business model. The results are discussed in Section 5, which provides insights on the second research question. We also outline managerial and policy implications of these results, pertaining to the third research question. Section 6 summarizes the key insights and provides an outlook on future research.

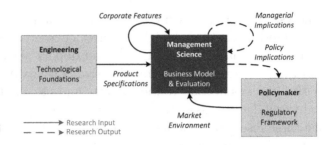

Figure IV–1: Research framework

2 Research Framework and Related Work

As illustrated in Figure IV–1, the primary objective of our research seeks to inform managers and policymakers on the potential of using electric vehicles for large-scale energy storage. We construct our study based on input from three areas. First, engineering contributes through the technological foundations of electric mobility and vehicle-to-grid (V2G) concepts. These describe the physical characteristics of the product the business model is built on. Second, the regulatory framework determined by policymakers shapes the market environment the business model would need to thrive in. Third, we also need to take corporate features of the business that seeks to implement EV4ES into account, such as existing business models within the corporation that may complement or interfere with EV4ES.

Our review of related literature covers these aspects by first illuminating the relationship between electric mobility and renewable energies. As a next step, we consider publications on the aggregation of electric vehicles that describe technological foundations and appropriate energy markets. We summarize work that suggests to use parking garage operators as intermediaries in EV4ES schemes, before we conclude this section by presenting seminal research on business models and outline concepts that relate to the research objectives of this paper.

2.1 Electric Mobility and Renewable Energy

During the past decade, the transportation sector has been under-going a steady but, nonetheless, radical shift as combustion engines are increasingly replaced by electric motors. This development is particularly evident for privately-owned vehicles, where a switch to electric mobility is often subsidized by the government. The reasons for the appeal of electric mobility can be found in decreased emissions of pollutants and noise, resulting in an improved quality of life in urban centers around the world. Naturally, the increasing electrification of mobility, along with the necessary preconditions and resulting impacts, have been accompanied by research efforts in a wide variety of disciplines. These include engineering (e.g. Lopes et al., 2011; Clement-Nyns et al., 2010; Sortomme et al., 2011; Tomić & Kempton, 2007; Saxena et al., 2015), management and operations research (e.g. Mak et al., 2013; Avci et al., 2014; Glerum et al., 2014; Chung & Kwon, 2015; Flath et al., 2014), as well as policy research (e.g. Galus et al., 2012; Lemoine et al., 2008; Andersen et al., 2009; Sovacool & Hirsh, 2009).

In the elaborations on electric mobility, the link to renewable energy sources is particularly noteworthy. On the one hand, electric vehicles only substantially contribute to a reduction in CO_2 emissions if the energy used to power the vehicles is largely generated by renewables (Faria et al., 2012, 2013; He & Chen, 2013). On the other hand, many of these energy sources, such as solar and wind power, are subject to exogenous factors and intermittent generation. As a result, they are difficult to align with energy demand. Intelligently charging and discharging plugged-in electric vehicles may contribute to a solution of this problem by using them as energy storage devices (Richardson, 2013; Hodge et al., 2010; Ekman, 2011). Furthermore, providing grid services may open additional revenue streams to EV owners that help compensate the steep cost difference between conventional and electric vehicles (Lave & MacLean, 2002; Delucchi & Lipman, 2001; Peterson et al., 2010). These synergies between electric mobility and renewable energy generation (Brandt et al., 2013) have resulted in extensive research on EV4ES concepts.

2.2 Electric Vehicle Aggregation and Frequency Regulation

EV4ES generally assumes that the storage capacity of multiple electric vehicles is aggregated by some intermediary entity and offered on an energy market. One of the first pilot projects on this topic was realized by Brooks (2002). He found that EVs are well suited for frequency regulation due to their short ramp-up time and negligible costs during idleness. Frequency control (or regulation) is a perpetual process that seeks to negate demand or supply shocks in the power system. It also describes an energy market where reserve energy used for this process is traded. Brooks (2002) calculates annual gross revenues per EV from USD 1,000 to USD 5,000, depending on individual driving activities.

Kempton & Tomić (2005a,b) build upon Brooks' work and lay the theoretical and conceptual foundations of the work on EV4ES developed during the following decade. They particularly focus on V2G-applications, which allow a bidirectional energy flow between the power grid and electric vehicles. In a study of Californian energy markets, they identify the markets for frequency regulation and spinning reserves as most profitable. However, they note that they would be saturated if just three percent of the vehicle fleet in California would switch to electric and offer their storage capacity to those markets. Given current adoption rates of electric mobility, frequency control is nonetheless still the most appealing market and will be the focus of this paper.

In a pilot study, Kempton et al. (2008) use a single EV to support frequency regulation in the PJM market. They find that V2G-capable electric vehicles can provide valuable ancillary services. This is confirmed by Kamboj et al. (2011), who further highlight the short response time of EVs as the main advantage over traditional regulation mechanisms. They estimate annual revenues ranging between USD 1,200 and USD 2,400 per vehicle. However, they assume that vehicles participate in regulation for 15 hours a day on average and set the price of regulation energy to be twice the normal value. The impact of these assumptions becomes evident when contrasted to the results from the *Mini E Berlin* field test (Vattenfall Europe Innovation GmbH et al., 2011). In this study, a fleet of 50 EVs is investigated, with the authors assuming that vehicles only contribute to regulation during 4 hours a day. The calculated revenues are at EUR 34 (about USD 40) per vehicle and year substantially lower.

Another crucial restriction is outlined by Kamboj et al. (2010), who note that market participation rules usually require the reliable ability to supply a certain amount of power, which is far beyond what a single EV can manage. Hence, a sufficient number of EVs have to be aggregated by an operating instance to participate in the market for frequency regulation. Quinn et al. (2010) also stress the importance of a mediating entity in such an aggregation scheme, particularly to facilitate communication between the *Independent System Operator* (ISO)—usually responsible for managing frequency regulation—and the electric vehicles.

2.3 Parking Garage Operators as Intermediaries

Possible intermediaries that are repeatedly suggested in the literature are operators of parking garages. For instance, Tulpule et al. (2011) emphasize the impact of charging multiple EVs with photovoltaic energy at workplace parking garages on power plants, transmission and distribution lines, as well as emissions. Chen et al. (2012, 2013) formulate the charging of vehicles in a large parking facility as a deadline scheduling problem, taking into account deterministic arrival, departure, and charging characteristics. Similarly, Sanchez-Martin & Sanchez (2011) implement a management system that handles battery charging of multiple parked EVs while taking the arrival and departure times into account. Kulshrestha et al. (2009) use an event-driven simulation to evaluate an energy management system that seeks to ensure optimal usage of available power, minimize charging time, and improve grid stability.

While parking garages appear to be appropriate as an intermediary for EV4ES, previous studies have largely focused on technical and operational aspects. Nevertheless, parking garage operators are generally for-profit corporations and EV4ES needs to prove itself as a viable business strategy under real-world conditions. We address this research gap by using literature on business models to investigate the business case and evaluate it using extensive real-world data sets.

2.4 Business Models

Research on business models has intensified following the ongoing digitalization during the past decade and the challenges many traditional models face (Alt & Zimmermann, 2014). From a corporate perspective, understanding a firm's existing business model is a crucial requirement to reinvent it and generate breakthrough innovations (Johnson et al., 2008). From an academic perspective, it is of particular interest to investigate why certain business models succeed or fail (Osterwalder, 2004). Veit et al. (2014) summarize the contribution of research on business models as "depicting, innovating and evaluating business logics in startups and in existing organizations" (p. 46).

In this paper, we build our arguments on concepts from Osterwalder (2004), which is generally considered to be one of the central publications of business model research (Alt & Zimmermann, 2014). We particularly focus on the building blocks, which are a synthesis of a comprehensive literature review on business models.

Furthermore, Abdelkafi et al. (2013) puts a particular emphasis on the business ecosystem surrounding electric mobility. While the proposed business model framework is only a slight extension of Osterwalder's building blocks, the context of electric vehicles gives the paper a unique perspective relevant to our case. In the next section, we will refer to these publications to dissect the business case for using parking garage operators as an intermediary in an EV4ES scheme.

3 Modelling the Business Case

At first glance, the idea to employ electric vehicles located at large parking facilities to support grid energy storage appears sound. After all, vehicles are parked during 95 to 97 percent of the day on average (Bates & Leibling, 2012; Shoup, 2005). Especially in urban areas, which EVs are particularly suited for due to relatively short distances, a sizable share of this parking demand will be covered by large garages. However, it is necessary to consider that parking is a multi-billion dollar service industry. For instance, in 2013 the two leading parking providers in Europe, Q-Park and APCOA, each administered several hundred thousand parking spots and generated revenues exceeding EUR 600 million (APCOA Parking Group, 2014; Q-Park, 2014). For these companies, EV4ES is first and foremost an economic decision. In this section, we analyze how well EV4ES aligns with the general business interests of parking garage operators, which synergies can be exploited, and which additional costs arise.

Osterwalder (2004) splits the foundations of a business model into four pillars – *product, customer interface, infrastructure management,* and *financial aspects.* He divides these further into building blocks, which will serve as the basis for the subsequent analysis.

3.1 Product

According to Osterwalder (2004), the product essentially describes what a firm offers to its customers. This is captured in the sole building block of this pillar—the *value proposition.* Following Kambil

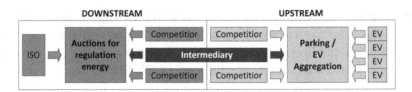

Figure IV–2: Market position of parking garage operator (intermediary)

et al., the value proposition "defines how items of value (product and service features as well as complementary services) are packaged and offered to fulfill customer needs" (Kambil et al., 1996, p. 6). To understand the value proposition for EV4ES, we need to be aware of the position of the parking garage operator within the market. This position is visualized in Figure IV–2.

EV4ES suggests that electric vehicles are aggregated and that their storage capacity is sold at an energy market. Hence, the parking garage operator is essentially an intermediary[1] that connects two different kinds of customers. In each market they face competitors, although a competitor in one market does not necessarily have to be a competitor in the other. Also, the intermediary needs to provide a value proposition to customers in the respective market. Taking energy storage as the "core product", we refer to the parking (and charging) market as the *upstream market* and the market for regulation energy as the *downstream market*.

Competitors in the market for regulation energy are mostly gas and pumped hydroelectric power plants that can supply energy quickly. The marketplaces are usually organized by Independent System Operators that are legally required to handle frequency disturbances in the grid. The energy used to balance these disturbances is acquired through auctions. With the rising share of intermittent renewable generation in the energy portfolio, the amount of reserve energy necessary for frequency regulation is expected to increase in the future

---

[1]In the remainder of this paper we use these terms interchangeably.

(Lund, 2007). Matching supply to this increase in demand would require substantial investments (e.g. gas turbines), associated with enormous upfront costs and uncertain revenues. The intermediary as an aggregator, but not owner, of electric vehicles possesses a competitive advantage in this scenario since necessary investments are much lower when compared to building new power plants.

Hence, the value proposition for the downstream market is the ability to provide regulation energy at relatively low fixed costs. Marginal costs are more difficult to assess. Consider negative regulation, where the grid frequency is too high and demand needs to be increased or supply curtailed. In this case, marginal costs are close to zero. The intermediary can simply increase the number of vehicles that are currently being charged and can in turn bill this energy to the vehicles' owners. However, this requires that not all vehicles are immediately charged upon entering the garage and connecting to a charge point (CP). For positive regulation, needed when the grid frequency is too low, the intermediary can reduce the number of vehicles being charged, as long as this does not interfere with any agreements made with the owners. Vehicles could also be discharged (V2G); however, in that case, owners would need to be compensated for the energy loss.

This relationship between the intermediary and vehicle owners is the core of the value proposition at the upstream market. The parking garage operator needs to attract a sufficient number of drivers of electric vehicles to be able to provide frequency regulation. Thus, the intermediary competes against other parking providers, as well as alternate means of transport, such as buses and trams. The first part of the value proposition to these target customers is simply the ability to charge the vehicle. Commercial parking providers have been slow to upgrade their facilities with charging stations. Hence, providing charging options is a simple way to stand out from the competition and be prepared for future needs. However, more crucial to the EV4ES business model is a share in the revenues from frequency regulation that needs to be passed on to the vehicles owners as incentives to participate. This can be realized through free or cost-reduced charging (negative regulation), payment for energy that has been discharged through V2G (positive regulation), or reduced parking fees.

3.2 Customer Interface

The second pillar of Osterwalder's business model ontology, customer interface, contains the firm's *target customers*, the *distribution channels* through which firm and customers interact, as well as the type of *relationship* the company intends to build with its customers. Essentially, these building blocks outline how and to whom a firm delivers its value proposition (Osterwalder, 2004).

We have previously mentioned that there are two types of customer, one for each market. The downstream market is usually heavily regulated and monopsonistic. For instance, the German market for regulation energy is an Internet platform operated by the four ISOs in Germany (regelleistung.net, 2013). These ISOs act as a single customer. Furthermore, the regulator must approve potential sellers of regulation energy before they are admitted into the market.

The target customers in the upstream market are owners of electric vehicles who park at the intermediary's garages. They may include drivers who park at the respective garage only once (e.g. during a business trip) or who enter a recurring relationship with the intermediary. The latter type of customer might use the parking garage on a daily basis to, for instance, park the vehicle during working hours. Depending on the overall composition of the customer base, the intermediary may need to tailor the product such that it appeals to either type. For instance, recurring customers may enter into a long-term contract, while one-time customers are offered the EV4ES product on a single-use basis. The intermediary can build upon existing billing infrastructure in the parking garages in either case. However, hardware and software will need to be adapted to provide and process EV4ES.

3.3 Infrastructure Management

The pillar *infrastructure management* in Osterwalder's (2004) ontology outlines how the value system is configured in order to deliver the value proposition and to maintain customer interfaces. It contains the building blocks *capabilities*, *value configuration*, and *partnerships*.

The capability concept relates to the *value creation* concept in Abdelkafi et al. (2013) as a resource transformation. Capabilities are repeatable uses of a firm's assets and resources to generate products or services (Osterwalder, 2004). The core competence of an EV4ES intermediary centers around the ability to aggregate the storage capacities of electric vehicles. This is reflected in two capabilities—*charging* and *scheduling*. Charging refers to the task of moving energy between the grid and the vehicle battery. It requires the necessary charging infrastructure, which needs to be maintained and operated. Scheduling describes the control of these charging processes in a way that the requirements set by the customers on both markets are satisfied.

The value configuration of the business model follows directly from the coordinative function of the scheduling capability. The value configuration outlines how activities within the business model are linked to create value for the customer (Osterwalder, 2004), represented by the concepts of *value chain*, *value shop*, and *value network*. An EV4ES intermediary is a classical example for the latter. Value networks employ mediating technology to connect different independent customers (Stabell & Fjeldstad, 1998). The intermediary *is not* the network, it rather *coordinates* the network. It brings EV owners and ISOs together for a joint gain.

Finally, the partnership building block describes strategic alliances between the intermediary and corporate partners. The idea is to join firm-specific capabilities for an improved value proposition. One possible partnership for the intermediary is to outsource the operation and maintenance of the charge points (CPs). This leaves the intermediary with the core capability of scheduling charging processes, as well as access to target customers and an existing distribution channel through the parking interface. Since outsourcing requires an extensive analysis of possible contractual models and exceeds the scope of this paper, we focus on the case in which the intermediary provides all necessary capabilities for the business model on its own.

3.4 Financial Aspects and Business Model Synopsis

Figure IV–3 summarizes the details of the business model we have discussed so far and leads to the final pillar. Financial aspects determine the overall profitability of the business model and directly follow from the composition of the other building blocks. A "firm's profit- or loss-making logic" (Osterwalder, 2004, p. 95) depends on its *cost structure* as well as its *revenue model.*

The cost structure of EV4ES includes initial investments into hardware and software infrastructure to enable the business model, as well as operating expenses to run the business model. We know that a core capability is charging the electric vehicles. Hence, the intermediary initially needs to build up the necessary charging infrastructure, captured in the cost account *CP Installation.* On the software side, the existing billing interface in the parking garages needs to be adapted to allow customers to charge their vehicle and participate in the EV4ES program. This holds for returning as well as one-time customers.

Operating expenses include the maintenance of charge points as well as the supervision and scheduling of charging processes. Major operating costs further result from the need to provide incentives to EV owners to participate in the EV4ES program. Hence, revenues need to be shared in some way, for instance, either by reducing the price of energy charged into the vehicles during negative regulation or by reducing parking fees. On the other hand, EV owners have to be compensated for energy transferred from the vehicle back to the grid during positive regulation. Similarly, prices for negative regulation at

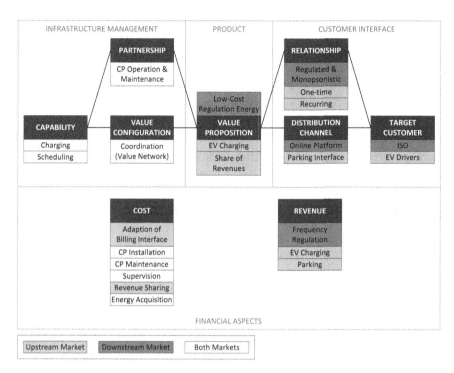

Figure IV–3: Business model for EV4ES by a parking garage operator

the downstream market do not necessarily have to be positive. Both aspects are captured in the cost account *Energy Acquisition*.

The revenue streams are threefold. First, the intermediary sells regulation energy at the downstream market. This is the main source of income directly related to the EV4ES concept. Second, parked vehicles can be billed for energy charged to their batteries. However, charging batteries is only directly related to EV4ES if it is done in response to a request for frequency regulation. Third, the intermediary as a parking garage operator receives fees from parked vehicles. However, these parking fees are only attributable to EV4ES if EV owners decided to park at the intermediary's facility in response to the EV4ES value proposition.

In summary, aggregating electric vehicles through parking garages is certainly a promising business case from a modeling perspective. We have a singular entity that has (albeit indirect) access to thousands of vehicles. A hardware and software interface to interact with the owners of these vehicles is already in place, although it would still need to be substantially adapted. In the next section, we will use data on parking garage occupation rates and regulation markets to further explore the profit-making logic of this business model.

4 Evaluation of the Business Model

Due to the infancy of real-world EV4ES implementations, it is not possible to directly observe the profitability of the business model proposed in this paper. Instead, we rely on real-world data sources that provide information on how such a business model would function. We first outline the energy market used in our study, the market for frequency regulation in Germany. Second, we describe the data sources our analysis mainly relies on—data on markets for frequency regulation, data on parking garage occupancy, and data on the states of charge of electric vehicles throughout the course of the day. Third, we describe the methods used to join all three data sources to estimate the upper bound of the main revenue streams in the business model. Fourth, we present the results from this analysis. We juxtapose these estimates to other cost and revenue sources in the business model to generate insights on its prospective profitability in the subsequent section.

4.1 Market for Frequency Regulation

Electronic markets for frequency regulation have been formed in most ISO regions in the U.S. (e.g. New England, New York, Texas, and California), Canada (e.g. Ontario, Alberta), and in various European countries like Germany, Switzerland, and Denmark (Kirby, 2004). We focus on the unified platform of the German Control Reserve Market (GCRM), as it coordinates frequency regulation for an area populated by more than 80 million people, making it one of the largest platforms in the world. However, most other electronic markets only differ in small details.

The GCRM operates daily auctions for primary, secondary, and tertiary control reserve, which most prominently differ in the respective ramp-up time (ENTSO-E, 2012). We focus on tertiary control reserves, also known as minute reserves, as this is the most expensive kind and well suited for EV aggregation by allowing sufficient time to coordinate the EV response. Market participants must be able to supply or absorb a minimum of 5 megawatts (MW) of power over a 4 hour interval with a ramp-up time of less than 15 minutes. Taking the example of conventional power outlets (230 volts, 16 ampere), this would require about 1,400 electric vehicles to concurrently (dis)charge. Fast-charging technology decreases this number substantially, although batteries are depleted much quicker.

Information on the market design and historic market data is available at the GCRM website[2]. Figure IV–4 depicts all auctions for a specific day and a specific frequency control region. The day is split into 6 segments of 4 hours each with one auction for positive energy, as well as for negative energy, for each of these segments. This results in a total of 12 auctions per day. Positive regulation energy implies supplying energy to the grid, either by generating energy or by reducing the load and is required if the frequency is below 49.8 Hz. By contrast, negative energy implies reducing energy generation or adding load and is required if the grid frequency exceeds 50.2 Hz.

Each auction is a particular kind of combinatorial reverse auction with asks being represented as $\{P, p_S, p_W\}$, i.e. the offered amount of power, the service price, and the work price. Naturally, asks must also carry information on which auction they are for, which is implicitly assumed as given for brevity. Power is the amount of electric power

---

[2]https://www.regelleistung.net/, accessed on April 13, 2015.

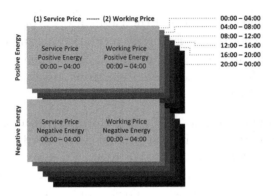

Figure IV–4: Daily auctions for tertiary control reserve on the GCRM

that can be supplied on demand, while the service price is the ask price for keeping this power available and is expressed in EUR per MW. The working price on the other hand is the ask price for energy that is actually supplied and is expressed in EUR per megawatt-hour (MWh).

The ISO determines a certain target quantity of power that it must acquire and accepts asks of increasing service prices until that amount is reached (or exceeded, since asks cannot be split). This target quantity varies for each day and auction and is substantially higher than the energy actually required for regulation to minimize risk. It is publicly announced by the ISO beforehand. The highest service prices accepted in previous auctions are public knowledge, as well.

The sellers are immediately paid the service compensation $R_S = P \cdot p_S$ if their offer is accepted. All accepted asks are then ranked in a merit order list with ascending working prices. Once regulation energy is actually required, offers for energy supply are accepted according to this list, with the cheapest suppliers first. Providers are paid the working compensation, R_W, which we express according to the formula

$$R_W = p_W \int_{t_0}^{t_1} P_a(t)dt \qquad (IV\text{--}1)$$

with $P_a(t)$ as the actual power supplied at t between the start, t_0, and the end, t_1, of the time span during which a regulation action is required, and $P_a(t) \leq P$.

To calculate an estimate of the upper bound of revenues that the intermediary could earn in a given year, we determine the critical prices $\overline{p_S}$ and $\overline{p_W}$ for each auction. $\overline{p_S}$ is the service price that was just still accepted, i.e. the marginal cost of capacity as incurred by the ISO. If there were one or multiple regulation actions within an auctioned time segment, $\overline{p_W}$ is defined as the working price that received the highest payout over all regulation actions in that time segment.

As an example, consider two suppliers of regulation energy, one holding a reserve capacity of 5 MW at EUR 1 per MWh supplied, and the other a reserve capacity of 5 MW at EUR 100 per MWh supplied. Assume that there are two regulation actions within the 4-hour time span that was auctioned off. First, consider the case where each regulation action required 5 MW of power over 15 minutes intervals. $\overline{p_W}$ would be 1 and the revenue would be

$$R_W = 1 \frac{\text{EUR}}{\text{MWh}} \cdot 5\,\text{MW} \cdot 2 \cdot 0.25\,\text{h} = 2.5\text{EUR}. \qquad \text{(IV--2)}$$

Now consider the case where the second regulation action requires 5.2 MW of power over 15 minutes. In that case $\overline{p_W}$ would be 100 EUR and revenue would equal

$$R_W = 100 \frac{\text{EUR}}{\text{MWh}} \cdot 0.2\,\text{MW} \cdot 0.25\,\text{h} = 5\text{EUR} \qquad \text{(IV--3)}$$

since the second supplier earns more than the first, even though only supplying 0.2 MW over 15 minutes. Hence,

$$\overline{p_W} = \arg\max_{p_W \in P_W} \left[p_W \int_{t_0}^{t_1} P_a(t)dt \right] \qquad \text{(IV--4)}$$

with P_W as the set of all observed working prices, as well as t_0 and t_1 as the start and end times of the auctioned time span, respectively.

Essentially, we assume that the intermediary always bids the perfect price. This will clearly not be the case in a real-world implementation, but provides us with an upper bound on regulation revenues that we can compare to the costs of the business model. Furthermore, we will be able to assess the profitability of the business model if only fractions of these revenues can be realized.

4.2 Data Sources

Data on the market for frequency regulation was acquired through the website of the GCRM. This includes data on all offered service prices, all accepted service prices, and the associated working prices. Furthermore, it contains information on regulation actions and the amount of regulation energy delivered.

Figure IV–5 illustrates core indicators of this data set. We observe that the service prices for negative regulation energy are much higher than for positive regulation both across the various time slots (Figure IV–5a) and throughout the year (Figure IV–5b). This suggests that the market for negative regulation is less saturated or that it is simply cheaper to keep measures for positive regulation available. The opposite phenomenon occurs when we consider working prices. Figure IV–5c outlines the average $\overline{pw}$-values for different auction types, i.e. the highest working prices that were paid when reserves were needed. Figure IV–5d depicts the same over the course of a year for the segment between noon and 4 p.m. In both diagrams prices for positive reserves far exceed those for negative reserves. The reason is that negative regulation energy can be used in production processes or resold. This provides the supplier of negative regulation with reductions in operational costs or an additional stream of revenues. Figure IV–5d also illustrates the rarity of regulation incidents during the year.

A crucial piece of information for the EV4ES business model is the *state of charge* (SoC) of the batteries when the electric vehicles enter

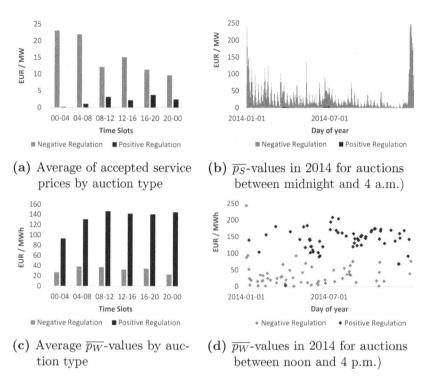

(a) Average of accepted service prices by auction type

(b) $\overline{p_S}$-values in 2014 for auctions between midnight and 4 a.m.)

(c) Average $\overline{p_W}$-values by auction type

(d) $\overline{p_W}$-values in 2014 for auctions between noon and 4 p.m.)

Figure IV–5: Prices at the GCRM for tertiary reserves in 2014

the parking garage. A high SoC decreases the immediate potential for negative regulation while increasing the immediate potential for positive regulation. We estimate the state of charge using data from privately used EVs in the Netherlands and Germany. Overall, the data set contains 34 million observations gathered across multiple years. Each observation consists of information on the state of charge of the battery, the state of the vehicle (driving, parking, charging, idle), as well as the type of vehicle (e.g. Nissan Leaf or Tesla Roadster). From these observations we derive distributions on the probability of certain SoCs during the day. Figure IV–6a illustrates the means of these distributions. We observe that, on average, the SoC of the observed vehicles varies between 60 and 80 percent throughout the day. Hence,

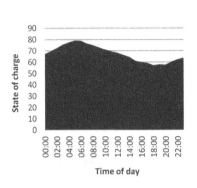

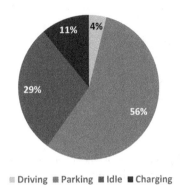

(a) Average state of charge at specific times of day

(b) Average shares of vehicle states during day

Figure IV–6: Data on electric vehicles

there is potential for negative as well as positive regulation. Figure IV–6b further confirms this assessment. Vehicles are charging for 11 percent of the day, while idling—meaning that they are connected to the grid but not charging due to a full battery—for 29 percent. This outlines that vehicles are connected to the grid for much longer time spans than necessary for charging—time that can be used for regulation actions. Figure IV–6b further shows that our vehicles are representative in the sense that the time spent driving (4 percent) is in line with the numbers found in large-scale studies mentioned earlier.

To model occupancy rates, we have collected data on the occupancy of 37 parking facilities (containing 15,000 parking slots) in the German cities of Freiburg and Frankfurt. We accessed the respective parking guidance systems of each city in 15 minutes intervals over several months to derive robust parking distributions. Figure IV–7 shows average occupancy during the day for two sample parking facilities in Freiburg. We observe that in both cases occupancy increases during the morning, peaks around noon, and decreases in the afternoon.

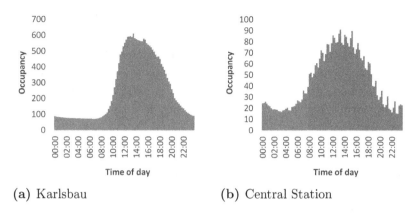

(a) Karlsbau   (b) Central Station

Figure IV–7: Occupancy of two sample parking garages in Freiburg

However, the shapes also differ due to the difference in sizes of the garages, as well as different durations of parking. The central station garage exhibits a more volatile pattern due to shorter parking durations. From this data, we can model the proportion of vehicles to expect at specific times of day, as well as the probabilities of specific parking durations (assuming a Chi-square distribution).

4.3 Modeling Annual Revenue

We estimate annual revenue by employing a computational experiment (e.g. Kydland & Prescott, 1996; Nicolaisen et al., 2001; Bichler et al., 2009) in combination with simulation methods. More specifically, for a given number N of vehicles entering the intermediary's parking facilities during the day, we calculate how much regulation energy can be provided as well as the upper bound of associated revenues (c.f. Section 4.1). The time when each vehicle arrives and the duration of its stay are drawn from the distributions derived in the previous subsection.

Once EVs are parked at the parking garage and plugged in, the intermediary needs to employ a scheduling strategy to determine when each vehicle charges or provides regulation energy. If all connected vehicles would supply regulation energy simultaneously, the amount requested by the ISO might be exceeded or circuits might be over-loaded. We have analyzed the monetary effects of different charging strategies in our previous work (Wagner et al., 2015). Since differences between strategies were minor, we employ a simple first-come-first-served strategy for this analysis.

Similar to Quinn et al. (2012), we model each time an EV is connected to a charge point in the garage as a *finite-state machine* (Arbib, 1969). Specifically, we define each parking event as a 5-tuple $v_n = (\Sigma, S, s_0, F, \delta)$ with

$$\Sigma = \{i_0, i_1, i_2, i_3, i_4\},$$
$$S = \{\text{idle}, \text{charge}, \text{up}, \text{down}, \text{drive}\},$$
$$s_0 = \text{idle},$$
$$F = \{\text{drive}\},$$
$$\delta : S \times \Sigma \rightarrow S.$$

Σ is the set of inputs to the machine, while S is the set of states of the machine. When a machine is initialized, i.e. an electric vehicle is being connected to a charge point, the initial state (s_0) is *idle*. The vehicle is just connected to the grid, but no (dis)charging occurs. The state *charge* refers to normal charging of the vehicle outside of any EV4ES arrangement. State *up* and *down* refer to positive and negative regulation, respectively (discharging and charging the vehicle). The state *drive* implies that the EV is disconnecting and leaving the facility. It is the only final state of the machine. The elements in Σ correspond to the elements in S, i.e. i_0 switches a vehicle to *idle*, i_1 switches it to *charge*, etc. The transition function δ maps the inputs and states onto the set of states. Transitions between all states are possible, with the exception of a transition from *charge* to *down* and vice versa.

Using this setup, we calculate the annual revenue from frequency regulation at the GCRM in 2014. We assume that the parking facilities are equipped with 25 kW fast charging stations and that the intermediary always offers 5 MW of reserve power, irrespective of whether the number of parked vehicles is sufficient to provide this power. We will discuss the implications of this assumption in Section 5.

4.4 Results

Our analysis seeks to answer two questions—how well the intermediary can supply regulation energy and what the financial payoff is. Figure IV–8 outlines the results pertaining to the first question. On the one hand, it depicts the number of electric vehicles that are concurrently parked at the facilities for a total of 10,000 vehicles entering the parking garages throughout the day. On the other hand, it shows how reliable these vehicles can supply 5 megawatts of negative regulation power. An interval length of 15 minutes is used for both variables.

We observe that the intermediary can reliably supply the required amount of power between 10 a.m. and 6 p.m. as marked by the dashed lines in Figure IV–8. However, in the early morning, evening, and at night, the number of vehicles entering the facilities in not sufficient to supply 5 megawatts for extended periods of time. Figure IV–8 also shows that, given our distributions on how long vehicles stay at the facility, about 800 concurrent EVs are necessary to reliable supply 5 megawatts of negative reserve. Naturally, a higher number of EVs throughout the day would improve reliability in the morning and afternoon. Yet, even 10,000 vehicles is an optimistic number, given current adoption rates of electric mobility.

The resulting estimates on annual revenue are shown in Table IV–1. We present two different scenarios. In the first case, *Full day*, the intermediary offers five megawatts of power during the entire day— even though, as shown in Figure IV–8, it is not able to supply this

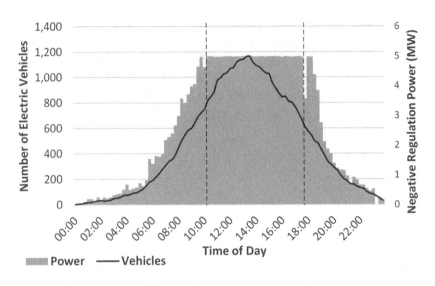

Figure IV–8: Average number of vehicles available and maximum
amount of negative regulation power supplied (Simulation
with 10,000 vehicles per day)

power during two thirds of the day. Assuming that the intermediary
would still be able to sell the energy it can supply, annual revenues are
EUR 510,656. In the second and more realistic case, the intermediary
only participates in the auctions for the *8 a.m. to noon* and *noon to
4 p.m.* segments (cf. Figure IV–4), during which a reliable supply of
five megawatts is more realistic. While these two segments constitute
a third of the day, revenues are decreased by less than half to EUR
274,992 since most regulation actions occur during this timespan.

Negative regulation energy only accounts for 7.0 percent of these total
annual revenues. However, negative regulation implies that load is
increased—vehicles are charged—and that this energy can be resold to
the vehicle owners. For this we assumed a price of 0.19 EUR per kWh.
Given average residential electricity prices of 0.30 EUR per kWh in

Table IV–1: Revenue evaluation (all values in EUR)

Revenue source	Full day	8 a.m. to 4 p.m.
Negative regulation (downstream)	26,232	19,125 (7.0%)
Negative regulation (upstream)	144,068	104,490 (38.0%)
Positive regulation	145,751	91,474 (33.3%)
Negative service	(170,905)	50,001 (18.2%)
Positive service	(23,700)	9,902 (3.6%)
Total	**510,656**	**274,992 (100.0%)**

Germany in 2014[3], this also provides an incentive to EV owners to participate in the regulation service. Energy resales in the upstream market account for EUR 104,490 or 38.0 percent of total revenues. Furthermore, positive regulation energy yields EUR 91,474 or 33.3% over the year. All positive regulation was provided by suspending charging processes of vehicles. No EV was discharged during positive regulation since prices were too low throughout the year[4]. The service prices for these two auction segments yielded EUR 50,001 and EUR 9,902, respectively. Revenues from positive and negative regulation both reflect the average price differences shown in Figure IV–5.

Recall that these numbers constitute an upper bound, at least for the downstream market. They results from the assumption that the intermediary always offers the exact price that is just still accepted by the ISO. On the other hand, the difference between the resale price of EUR 0.19 and the average electricity price of EUR 0.30 is quite large. Raising the resale price to EUR 0.25 would increase the

---

[3]http://appsso.eurostat.ec.europa.eu/nui/show.do?dataset=nrg_pc _204&lang=en, accessed on April 13, 2015.
[4]Vehicles would only be discharged if the price for positive regulation exceeds the average residential electricity price in Germany of EUR 0.30.

revenues from this account to EUR 137,486. Hence, it might be worth considering to completely focus on resale revenues and set the service and working prices for negative energy to zero (such that offers are always accepted). Together with yields from positive regulation, the intermediary can realistically expect annual revenues between EUR 150,000 and EUR 200,000.

5 Discussion

The revenue streams estimated in the previous section need to be put in relation to the cost structure of the business model. The main fixed cost factors are investments into charging infrastructure and the adaptation of the existing IT system for billing. Additionally, operations need to be supervised and charge points must be maintained (cf. Figure IV–3). The cost factors *Revenue Sharing* and *Energy Acquisition* are already accounted for. Revenues are shared by providing EV owners with cheap energy from negative regulation to charge their vehicles. EVs do not need to be discharged during positive regulation, which eliminates the need for energy acquisition.

The remaining cost factors need to be considered from two perspectives. The first requires the business model to stand on its own, in which case revenues need to recover investments within a reasonable timespan. For the second, it is part of a broader business strategy that equips parking facilities with charge points, so that customers can charge their vehicles. In the latter case, we need to analyze how the EV4ES business model complements or interferes with this strategy.

The first case is quite straightforward. Just by considering the necessary investments into charging infrastructure it becomes evident that revenues are far inferior. Figure IV–8 shows that around 800 vehicles need to be simultaneously connected to reliably supply five megawatts of regulation power. Even if we assume that new vehicles always park at the very garage in which a charge point just became available and that each charge point can accommodate two vehicles at a time, this would require 400 charge points. Given that each charge point costs approximately EUR 10,000 to install, necessary investments into charging infrastructure amount to four million euro. Annual revenues of EUR 200,000 or less cannot recover this within a reasonable timespan.

The second case is more complex. The parking garage operator may decide to equip its facilities with the necessary charging infrastructure and update its billing software to provide an additional service to its customers. EV owners are now able to charge their vehicles at the facility while they are parked there. Depending on the contractual structure, the intermediary may be able to buy energy in bulk at average industry prices of 0.21 EUR per kWh[5] and resell it to its customers at the residential price of 0.30 EUR per kWh, yielding a net profit of 0.09 EUR per kWh. Recall, though, that to participate in the market for negative regulation, the intermediary would need to keep five megawatts of power over four hours available for the case of a regulation incident. These twenty megawatt-hours represent vehicles that would be charged upon entering the garage. Hence, to participate in frequency regulation, the intermediary would forfeit selling 40 MWh of energy at an operating profit of 90 EUR per MWh for every day of the year. These opportunity costs add up to EUR 1.3 million per year, far outweighing possible revenues from frequency regulation.

These results have far-reaching implications. On a managerial level, they outline the challenge of developing a business model for EV4ES. Parking garage operators are among the best candidates for EV4ES since they have access to a large number of vehicles and can partly rely on existing infrastructure. Yet, even in this case revenues are vastly inferior to necessary investments. Even if investments were covered by subsidies or a broader business strategy, simply selling energy

---

[5]http://appsso.eurostat.ec.europa.eu/nui/submitViewTableAction.do, accessed on April 13, 2015.

to EV owners results in substantially higher revenues than ancillary services such as frequency regulation. In addition, under current regulations the intermediary would not be approved to participate in the regulation market since it cannot absolutely reliably supply the necessary energy. The times at which vehicles enter and leave garages follow stochastic processes and even if the chance is slim, the number of vehicles may be insufficient to supply the required amount of power at the time of a regulation incident.

These facts have direct implications for policy. The current market environment does not allow EVs to provide frequency regulation. While this is not necessarily new information, our results show that the lack of an appropriate market is not the critical issue for the EV4ES feasibility. Even if a market for stochastic reserves existed, it would need energy prices of at least ten times the current level to result in a viable business model. Overall, while parking garage operators are certainly not the only possible intermediary, our results suggest that the potential of EV aggregators to provide ancillary services is generally exaggerated. The sole fact of being technologically feasible *does not* imply that it also constitutes an economically viable business model. Therefore, the widespread adoption of electric mobility may become a more critical issue for the power grid than is currently expected.

6 Conclusion

In this paper, we have developed and evaluated a business model for the aggregation of electric vehicles to provide energy storage. More specifically, we have analyzed whether such an aggregator can support the integration of intermittent renewable energy sources by providing energy for frequency regulation.

Based on similar approaches in the literature, we have identified parking garage operators as suitable candidates for this purpose. They have access to a large number of vehicles and an IT infrastructure for billing is already in place that could be adapted for the additional service. The parking garage operator would serve as an intermediary between the owners of electric vehicles and the market for frequency regulation. We use Osterwalder's (2004) business model ontology to further analyze the business model and identify the cost and revenue structure.

We simulate the main revenue streams using extensive real-world data sets on the respective market, occupancy rates of parking garages, and battery states of electric vehicles. When comparing the revenues to the necessary investment costs, we find that the former are vastly inferior to the latter. However, even if investment costs can be ignored, the opportunity costs arising from not immediately charging vehicles at the parking facilities outweigh possible revenues from frequency regulation. While previous studies have largely focused on the technological feasibility of aggregation concepts, we show that they are currently far from economic viability. In addition to the need for markets that allow stochastic reserves, the energy prices at these markets need to be ten times the current level for a viable business model to arise. Overall, the challenge, which a widespread adoption of electric mobility poses to the grid, may be fundamentally underestimated since economic factors are not sufficiently taken into account.

7 References

Abdelkafi, N., Makhotin, S. & Posselt, T. (2013). Business Model Innovations for Electric Mobility – What can be Learned from Existing Business Model Patterns? *International Journal of Innovation Management, 17*(01), 1340003-1-1340003-41.

Alt, R. & Zimmermann, H.-D. (2014). Status of business model and electronic market research: An interview with Alexander Osterwalder. *Electronic Markets*, *24*(4), 243–249.

Andersen, P. H., Mathews, J. A. & Rask, M. (2009). Integrating private transport into renewable energy policy: The strategy of creating intelligent recharging grids for electric vehicles. *Energy Policy*, *37*(7), 2481–2486.

APCOA Parking Group. (2014). *APCOA Parking to embark on new growth path*. Stuttgart/London. Retrieved on April 13, 2015, from `http://www.apcoa.com/press-news/read-article/apcoa-parking-to-embark-on-new-growth-path.html`

Arbib, M. A. (1969). *Theories of Abstract Automata*. Englewood Cliffs, NJ: Prentice-Hall.

Avci, B., Girotra, K. & Netessine, S. (2014). Electric Vehicles with a Battery Switching Station: Adoption and Environmental Impact. *Management Science*, *61*(4), 772–794.

Bates, J. & Leibling, D. (2012). *Spaced Out: Perspectives on parking policy*. London.

Bichler, M., Shabalin, P. & Pikovsky, A. (2009). A Computational Analysis of Linear Price Iterative Combinatorial Auction Formats. *Information Systems Research*, *20*(1), 33–59.

Brandt, T., Feuerriegel, S. & Neumann, D. (2013). Shaping a Sustainable Society: How Information Systems Utilize Hidden Synergies between Green Technologies. *ICIS 2013 Proceedings*, Paper 7.

Brooks, A. N. (2002). *Vehicle-to-Grid Demonstration Project: Grid Regulation Ancillary Service with a Battery Electric Vehicle*.

Chen, S., Ji, Y. & Tong, L. (2012). Large scale charging of Electric Vehicles. In *2012 IEEE Power & Energy Society General Meeting* (pp. 1–9). IEEE.

Chen, S., Mount, T. & Tong, L. (2013). Optimizing Operations for Large Scale Charging of Electric Vehicles. In *46th Hawaii International Conference on System Sciences* (pp. 2319–2326). IEEE.

Chung, S. H. & Kwon, C. (2015). Multi-period planning for electric car charging station locations: A case of Korean Expressways. *European Journal of Operational Research, 242*(2), 677–687.

Clement-Nyns, K., Haesen, E. & Driesen, J. (2010). The Impact of Charging Plug-In Hybrid Electric Vehicles on a Residential Distribution Grid. *IEEE Transactions on Power Systems, 25*(1), 371–380.

Delucchi, M. A. & Lipman, T. E. (2001). An analysis of the retail and lifecycle cost of battery-powered electric vehicles. *Transportation Research Part D: Transport and Environment, 6*(6), 371–404.

Ekman, C. K. (2011). On the synergy between large electric vehicle fleet and high wind penetration – An analysis of the Danish case. *Renewable Energy, 36*(2), 546–553.

ENTSO-E. (2012). *Continental Europe Operation Handbook.* Retrieved on April 13, 2015, from `https://www.entsoe.eu/publications/system-operations-reports/operation-handbook/Pages/default.aspx`

Faria, R., Marques, P., Moura, P., Freire, F., Delgado, J. & de Almeida, A. T. (2013). Impact of the electricity mix and use profile in the life-cycle assessment of electric vehicles. *Renewable and Sustainable Energy Reviews, 24*, 271–287.

Faria, R., Moura, P., Delgado, J. & de Almeida, A. T. (2012). A sustainability assessment of electric vehicles as a personal mobility system. *Energy Conversion and Management, 61*, 19–30.

Flath, C. M., Ilg, J. P., Gottwalt, S., Schmeck, H. & Weinhardt, C. (2014). Improving Electric Vehicle Charging Coordination Through Area Pricing. *Transportation Science, 48*(4), 619–634.

Galus, M. D., Waraich, R. A., Noembrini, F., Steurs, K., Georges, G., Boulouchos, K., ... Andersson, G. (2012). Integrating Power Systems, Transport Systems and Vehicle Technology for Electric Mobility Impact Assessment and Efficient Control. *IEEE Transactions on Smart Grid*, *3*(2), 934–949.

Glerum, A., Stankovikj, L., Thémans, M. & Bierlaire, M. (2014). Forecasting the Demand for Electric Vehicles: Accounting for Attitudes and Perceptions. *Transportation Science*, *48*(4), 483–499.

He, L.-Y. & Chen, Y. (2013). Thou shalt drive electric and hybrid vehicles: Scenario analysis on energy saving and emission mitigation for road transportation sector in China. *Transport Policy*, *25*, 30–40.

Hodge, B.-M. S., Huang, S., Shukla, A., Pekny, J. F. & Reklaitis, G. V. (2010). The Effects of Vehicle-to-Grid Systems on Wind Power Integration in California. In *20th European Symposium on Computer Aided Process Engineering* (Vol. 28, pp. 1039–1044). Elsevier.

Johnson, M. W., Christensen, C. M. & Kagermann, H. (2008). Reinventing Your Business Model. *Harvard Business Review*, *86*(12), 50–59.

Kambil, A., Ginsberg, A. & Block, M. (1996). Re-Inventing Value Propositions. *NYU Working Paper Series*, *2451/14205*.

Kamboj, S., Decker, K. S., Trnka, K., Pearre, N., Kern, C. & Kempton, W. (2010). Exploring the formation of Electric Vehicle Coalitions for Vehicle-To-Grid Power Regulation. In *First International Workshop on Agent Technologies for Energy Systems* (Vol. 1, pp. 1–8).

Kamboj, S., Kempton, W. & Decker, K. S. (2011). Deploying power grid-integrated electric vehicles as a multi-agent system. In *10th International Conference on Autonomous Agents and*

Multiagent Systems (Vol. 1, pp. 13–20). International Foundation for Autonomous Agents and Multiagent Systems.

Kempton, W. & Tomić, J. (2005a). Vehicle-to-grid power fundamentals: Calculating capacity and net revenue. *Journal of Power Sources, 144*(1), 268–279.

Kempton, W. & Tomić, J. (2005b). Vehicle-to-grid power implementation: From stabilizing the grid to supporting large-scale renewable energy. *Journal of Power Sources, 144*(1), 280–294.

Kempton, W., Udo, V., Huber, K., Komara, K., Letendre, S., Baker, S., ... Pearre, N. (2008). *A Test of Vehicle-to-Grid (V2G) for Energy Storage and Frequency Regulation in the PJM System: Results from an Industry-University Research Partnership.*

Kirby, B. J. (2004). Frequency Regulation Basics and Trends. *Oak Ridge National Laboratory Technical Report, ORNL/TM-2004/291.*

Kulshrestha, P., Wang, L., Chow, M.-Y. & Lukic, S. (2009). Intelligent energy management system simulator for PHEVs at municipal parking deck in a smart grid environment. In *2009 IEEE Power & Energy Society General Meeting* (pp. 1–6). IEEE.

Kydland, F. E. & Prescott, E. C. (1996). The Computational Experiment: An Econometric Tool. *The Journal of Economic Perspectives, 10*(1), 69–85.

Lave, L. B. & MacLean, H. L. (2002). An environmental-economic evaluation of hybrid electric vehicles: Toyota's Prius vs. its conventional internal combustion engine Corolla. *Transportation Research Part D: Transport and Environment, 7*(2), 155–162.

Lemoine, D. M., Kammen, D. M. & Farrell, A. E. (2008). An innovation and policy agenda for commercially competitive plug-in hybrid electric vehicles. *Environmental Research Letters, 3*(1), 014003.

Lopes, J. A. P., Soares, F. J. & Almeida, P. M. R. (2011). Integration of Electric Vehicles in the Electric Power System. *Proceedings of the IEEE*, *99*(1), 168–183.

Lund, H. (2007). Renewable energy strategies for sustainable development. *Energy*, *32*(6), 912–919.

Mak, H.-Y., Rong, Y. & Shen, Z.-J. M. (2013). Infrastructure Planning for Electric Vehicles with Battery Swapping. *Management Science*, *59*(7), 1557–1575.

Nicolaisen, J., Petrov, V. & Tesfatsion, L. (2001). Market Power and Efficiency in a Computational Electricity Market With Discriminatory Double-Auction Pricing. *IEEE Transactions on Evolutionary Computing*, *5*(5), 504–523.

Osterwalder, A. (2004). *The business model ontology: A proposition in a design science approach*. Lausanne, Switzerland: University of Lausanne.

Peterson, S. B., Whitacre, J. F. & Apt, J. (2010). The economics of using plug-in hybrid electric vehicle battery packs for grid storage. *Journal of Power Sources*, *195*(8), 2377–2384.

Q-Park. (2014). *Annual Report 2013: Key Figures*. Retrieved on April 13, 2015, from `http://annualreport2013.q-park.com/about-q-park/key-figures`

Quinn, C., Zimmerle, D. & Bradley, T. H. (2010). The effect of communication architecture on the availability, reliability, and economics of plug-in hybrid electric vehicle-to-grid ancillary services. *Journal of Power Sources*, *195*(5), 1500–1509.

Quinn, C., Zimmerle, D. & Bradley, T. H. (2012). An Evaluation of State-of-Charge Limitations and Actuation Signal Energy Content on Plug-in Hybrid Electric Vehicle, Vehicle-to-Grid Reliability, and Economics. *IEEE Transactions on Smart Grid*, *3*(1), 483–491.

regelleistung.net. (2013). *FAQ for online platform.* Retrieved on April 13, 2015, from `https://www.regelleistung.net/ip/action/static/faq`

Richardson, D. B. (2013). Electric vehicles and the electric grid: A review of modeling approaches, Impacts, and renewable energy integration. *Renewable and Sustainable Energy Reviews, 19,* 247–254.

Sanchez-Martin, P. & Sanchez, G. (2011). Optimal electric vehicles consumption management at parking garages. In *2011 IEEE PowerTech* (pp. 1–7). IEEE.

Saxena, S., Le Floch, C., MacDonald, J. & Moura, S. (2015). Quantifying EV Battery End-of-Life through Analysis of Travel Needs with Vehicle Powertrain Models. *Journal of Power Sources, 282,* 265–276.

Shoup, D. C. (2005). *The High Cost of Free Parking.* Chicago: APA Planners Press.

Sortomme, E., Hindi, M. M., MacPherson, S. D. James & Venkata, S. S. (2011). Coordinated Charging of Plug-In Hybrid Electric Vehicles to Minimize Distribution System Losses. *IEEE Transactions on Smart Grid, 2*(1), 198–205.

Sovacool, B. K. & Hirsh, R. F. (2009). Beyond batteries: An examination of the benefits and barriers to plug-in hybrid electric vehicles (PHEVs) and a vehicle-to-grid (V2G) transition. *Energy Policy, 37*(3), 1095–1103.

Stabell, C. B. & Fjeldstad, Ø. D. (1998). Configuring value for competitive advantage: on chains, shops, and networks. *Strategic Management Journal, 19*(5), 413–437.

Tomić, J. & Kempton, W. (2007). Using fleets of electric-drive vehicles for grid support. *Journal of Power Sources, 168*(2), 459–468.

Tulpule, P., Marano, V., Yurkovich, S. & Rizzoni, G. (2011). Energy economic analysis of PV based charging station at workplace parking garage. In *2011 IEEE Energytech* (pp. 1–6). IEEE.

Vattenfall Europe Innovation GmbH, BMW AG, TU Berlin, TU Chemnitz & TU Ilmenau. (2011). *Gesteuertes Laden V2.0: Gemeinsamer Abschlussbericht.*

Veit, D., Clemons, E., Benlian, A., Buxmann, P., Hess, T., Kundisch, D., ... Spann, M. (2014). Business Models. *Business & Information Systems Engineering, 6*(1), 45–53.

Wagner, S., Brandt, T. & Neumann, D. (2015). IS-Centric Business Models for a Sustainable Economy – The Case of Electric Vehicles as Energy Storage. *WI 2015 Proceedings*, 1055–1070.

V Mechanism Stability—Flash Crashes and Avalanche Effects

This chapter is a revised and extended version of the papers:

Tobias Brandt and Dirk Neumann, "Chasing Lemmings: Modelling IT-induced Misperceptions about the Strategic Situation as a Reason for Flash Crashes" (2015), Journal of Management Information Systems, 31(4), 88–108.

Tobias Brandt and Dirk Neumann, "When Common Knowledge Becomes Common Doubt – Modeling IT-Induced Ambiguities about the Strategic Situation as Reasons for Flash Crashes" (2014), HICSS 2014 Proceedings, 1202–1211. Nominated for Best Paper Award.

Abstract

Flash crashes, perceived as sharp drops in market prices that rebound shortly after, have turned the public eye towards the vulnerability of IT-based stock trading. In this paper, we explain flash crashes as a result of actions made by rational agents. We argue that the advancement of information technology, which has long been associated with competitive advantages, may cause ambiguities with respect to the game form that give rise to a Hypergame. We employ Hypergame Theory to demonstrate that a market crash constitutes an equilibrium state if players misperceive the true game. Once the ambiguity is resolved, prices readjust to the appropriate level, creating the characteristic flash crash effect. By analyzing the interaction with herd behavior, we find that flash crashes may be unavoidable and a systemic problem of modern financial markets. Furthermore, we outline that flash-crash-like effects are also relevant in other applications that rely on increasing automation, such as the automated management of energy demand.

1 Introduction

On May 6, 2010, the effects of the "Flash Crash" rippled through U.S.-based equity markets and beyond. Within minutes, indices dropped by several percentage points, followed by an equally fast rebound (CFTC and SEC, 2010). In the immediate aftermath, several hypotheses that attempt to explain why the crash occurred have arisen. Easley et al. (2011) highlight some of the most prominent suggestions, for instance, a trader entering the wrong number for a transaction—a "fat-finger trade". While this has been quickly debunked, other possible causes include technical reporting difficulties at the NYSE, as well as currency movements between dollar and yen. Filimonov & Sornette (2012) also investigate the impact of herding (Devenow & Welch, 1996) on the likelihood of flash crashes. Several of these hypotheses assume irrational behavior or a lack of information on market conditions on behalf of the traders. However, they fail to take into account the effects of the tremendous advances in information technology (IT) that have been witnessed over the past decades and in recent years. In this paper, we show that flash crashes can be explained as a result of ambiguities regarding the strategic situation induced by this progress in information technology, despite rational and well-informed behavior. We also illustrate that this phenomenon does not contradict the influence of herding. Instead, the impact of these ambiguities amplifies and is amplified by the effects of herding.

Undoubtedly, the simultaneous selling frenzy of many traders is an unlikely candidate for a mutual best-response equilibrium. However, we argue that technological progress induces changes not only in the strategic environment, but also in the nature of the game, such that these flash crashes are best-response equilibria. This progress is best exemplified by the emergence of high-frequency trading and stock exchanges constantly having to upgrade their matching engines to ensure that trades can be matched at micro-seconds. It is also expressed on a broader scale by the replacement of human employees in the collection and processing of information by IT—processes vital to the functioning of financial markets. Our key assumption is that, in recent years, IT has eroded the competitive advantage dominant traders held over the rest. As the reaction time in financial markets dropped to milliseconds or microseconds, the game form in the market has been blurred. In some cases, it turned from a sequential game, with a leader informing the actions of other traders, towards a purely simultaneous game, where some or all traders literally trade at the same time. We investigate situations where traders do not recognize this shift in the strategic situation. This ambiguity violates the classic common knowledge assumption that all traders know the accurate strategic situation they are acting in. We model this situation by employing hypergame theory. This branch of game theory analyzes equilibria in cases where players perceive varying subjective games. As a result, flash crashes can be explained as best-response equilibria, with traders playing the dominant strategy in their subjective (sequential) game—what they perceive to be the true game—while in reality the game is simultaneous. Crashes occur when multiple traders act according to a sequential game or when the actions of a few traders are reinforced by herding.

While this paper uses financial markets as the primary showcase, the implications are much more far reaching. We can apply our approach to analyze any electronic coordination mechanism with sudden spikes in demand or supply. One such application is demand response, where households are given price signals so that they can decide when to use electricity. Demand response is intended to smooth overall energy consumption, since households will postpone some activities to later hours if the current price is high. However, if many households react similarly to a price signal, this may create unintended spikes in demand rather than removing them—so-called avalanche effects (Gottwalt et al., 2011). As an extension of our model, we explain the occurrence of such avalanches and discuss possible remedies in the penultimate chapter.

Consequently, the primary objectives of this paper are:

To analyze how the collection and processing of information influence the game form in a financial market and how they are affected by information technology;

To investigate the effects of misperceptions of the strategic situation on strategies and equilibria using hypergame theory;

To assess the combined effect of this misperception and herding as causes of flash crashes; and

To outline that flash crashes do not only endanger financial markets, but also other automated coordination mechanisms such as demand response.

2 Related Work

In this section, we review research that pertains to our work. This includes publications on flash crashes and market crashes in general, information processing and competitive advantages, as well as hypergame theory. We also detail the contribution of our research in the context of these works.

2.1 Market Crashes

The flash crash on May 6, 2010, caused extensive investigations by the U.S. Commodity Futures Trading Commission and the U.S. Securities and Exchange Commission (CFTC and SEC, 2010). They identified liquidity issues due to large sell orders as a possible reason for this crash. They inferred that the automated execution of those orders can induce extreme price movements. This effect is further amplified by interactions with algorithmic trading strategies. Easley et al. (2011, 2012) provide further evidence that liquidity problems were developing in the days before the event. They develop a metric to measure the toxicity of the order flow as an indicator of looming illiquidity. Kirilenko et al. (2011) emphasize the response of high frequency traders to selling pressure, which further amplifies market volatility. In a prediction model of flash crashes, Filimonov & Sornette (2012) highlight the relevance of exogenous information to market stability. They quantify the proportion of price movements that is due to exogenous news and demonstrate that this share has substantially decreased between 1998 and 2010. The endogenous part of the price movement reflects herd behavior in the market. The authors show that on May 6, abnormally high levels of endogeneity coincide with the flash crash, implying that "more than 95% of the trading was due to endogenous triggering effects rather than genuine news" (p. 8). This provides evidence for the contributing effect of herding to the flash crash. It is also in line with the fact that crashes in financial markets in general have been linked with herd behavior for several decades. For instance, Barlevy & Veronesi (2003) attribute herd behavior to rational but uninformed traders, while Lux (1995) proposes a cyclic model that explains bubbles and crashes as the result of such behavior.

The flash crash has also received attention within the Information System community. For instance, Lucas et al. (2013) pose the question as to what degree IT has contributed to the crash. In addition, Gomber & Haferkorn (2013), as well as Lattemann et al. (2012), investigate high-frequency and algorithmic trading in general and in connection with the flash crash.

2.2 Information Processing and Competitive Advantage

Our approach relates to research on information processing in organizations and on the competitive advantage induced by information technology, which are both established research fields in the Management Information Systems discipline. For instance, Premkumar et al. (2005) discuss the fit between information processing needs and capabilities in interorganizational supply chains. Wang et al. (2013) investigate the benefits to buyers and suppliers from information processing capabilities in supply chains. The revolutionizing effect of information processing capabilities in financial markets, which is a core component of our research, is captured in various publications on electronic markets, such as Lee & Clark (1996/1997). They investigate how IT adoption reduces transaction costs and increases market efficiency. Clemons & Weber (1990) discuss changes to the London Stock Exchange with the introduction of screen-based dealing systems and the competitive advantage that it conferred to the marketplace. Furthermore, Walczak (1999) analyzes how neural networks can equip traders with a competitive advantage in emerging capital markets. More recently, the relevance of information technology in today's markets is emphasized by Chlistalla (2011), who argues that the emergence of high-frequency traders was enabled by "significant advances in information technology" (p. 1). Zhang & Riordan (2011) point out that these technological innovations have substantially changed financial markets.

Budish et al. (2013) compare these changes to an arms race, with traders and exchanges constantly attempting to become faster. They also point out the challenges for the market design of exchanges resulting from this development. The "arms race" metaphor relates to a notion by Carr (2004), who questions the general sustainability of IT-induced competitive advantages. In the financial market as a market that is highly information sensitive, such a competitive advantage materializes as the ability to react faster to new information. The race for speed we face today has certainly been fuelled by the emergence of the Internet and the technologies surrounding it, from smartphones to social networks. They have fundamentally changed the way we gain access to and process new information—we can find real-time news on anything at any time basically anywhere (Sakaki et al., 2010; White, 2011). Therefore, we argue that it has become more and more difficult to create competitive advantages in such information-sensitive mechanisms, since agents often use similar information channels and process this information with similar software employing similar algorithms.

2.3 Game Form and Hypergame Theory

Unarguably, those agents who are best able to collect and process new information have always been at an advantage in financial markets by reacting first and making the best deals. The role of information arrival, i.e. whether agents receive news simultaneously or sequentially, has been researched for financial markets by Copeland & Friedman (1987). Groth (2010) analyzes the relationship between news-related liquidity shocks and automated trading engines. Xu & Zhang (2009) investigate how social media has changed the information environment in financial markets. Both studies show that information technology has fundamentally altered information dissemination, possibly changing the strategic situation from sequential to simultaneous.

We analyze what happens when agents are not aware of this game form ambiguity using Hypergame Theory, which was first proposed by Bennett (1977, 1980). It was originally used to model conflicts where each party has a different perception of the game being played. More recently, Sasaki & Kijima (2012) have compared hypergames with Bayesian games. While they show that most concepts from hypergame theory can be captured by a Bayesian game, they do point out that hypergames exhibit a persuasive intuitiveness. In our model, we exploit this intuitiveness to illustrate the effects of game misperception.

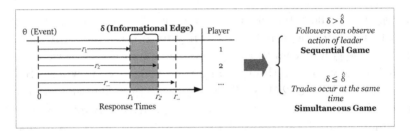

Figure V–1: Informational edge and game form determination

2.4 Contribution

While sudden liquidity erosion has been proposed as a possible cause of flash crashes, we investigate whether it can be considered to be a symptom of a more fundamental underlying strategic issue—a shift in the market environment caused by the advancement of information technology. We develop a game-theoretical model that explains flash crashes as the result of IT-induced ambiguities in the strategic situation of the traders. We analyze the effect of these ambiguities, both isolated and in association with herd behavior. We demonstrate the relevance of our results to other fields of research through an application to the automated management of energy demand.

3 The Role of Information in Game Form Determination

In this section, we make the argument for the central premise of this paper: *in some instances, information technology has changed the game that is played in financial markets from sequential to simultaneous.*

3.1 Sequential and Simultaneous Games

The form of the game fundamentally depends on each player's capacity to attain, process, and react upon new information. This relationship is illustrated in Figure V–1, which depicts players' responses to an event Θ. Such an event includes everything that might induce market activity—the publication of a company's annual report, sales figures of a newly released product, a major accident at a factory, or even the action of another player. Naturally, traders differ in their ability to collect and process information about an event. The sum of the time needed to receive information on the event, process this information, and come to a decision is captured in the response time r_i of each player $i \in I$. The decisive question is, whether the action of the fastest trader can be observed by other traders. If so, the other traders can include this action into their own decision-making process, resulting in a sequential Stackelberg game with one leader and $|I| - 1$ follower(s). Otherwise, the actions of at least two players are simultaneous. We formalize this by introducing the concept of the informational edge δ, which is essentially the competitive advantage the fastest trader holds over the rest of the field. Assuming that r_i follows an unknown distribution, the informational edge is the difference between the $|I|$th and $(|I| - 1)$th order statistics of r_i, i.e. $\delta = r_{(|I|-1)} - r_{(|I|)}$. However, for a sequential game to arise, δ needs to exceed a certain threshold value $\hat{\delta}$, which depends on the market mechanism. This threshold reflects the maximum amount of time at which the market mechanism still considers two actions to arrive simultaneously. It is determined by the information processing capabilities of the mechanism or by regulations.

To characterize the resulting games, we construct the following scenario. Since game form determination only depends on the $|I|$th and $(|I| - 1)$th order statistics, we can set $I = \{1, 2\}$ without a loss of generality. Both players are shareholders of a company trading at a stock exchange. The company is publishing its quarterly financial report, thus supplying shareholders with new information (the event

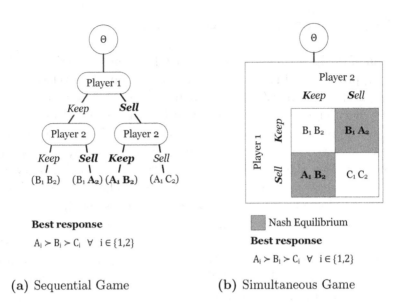

(a) Sequential Game   (b) Simultaneous Game

Figure V–2: Sequential and simultaneous game

Θ). We assume that shareholders have built expectations concerning this report in the preceding weeks and that these expectations are fully reflected in their respective portfolio. Each player makes a binary decision between selling and keeping a fixed share of its stock in the company and is aware of the payoff matrix of the game. The resulting games are visualized in Figure V–2.

Figure V–2a illustrates the sequential game, with payoffs expressing a preference structure. If both players keep their shares, they basically ignore the information from the event Θ. It is reasonable to expect that the value of their shares will decrease in the future compared to the price before publication of the report as the company struggles. If player 1 sells, they can get out before other market participants are aware of the report and prices are still reasonably high, thus utilizing the informational edge. The same applies to player 2 if player 1 did not sell. However, if player 2 sells after player 1 did so, the market has processed the action of player 1 and prices have dropped substantially. There is a clear subgame perfect Nash equilibrium within the game, which can be easily traced by backward induction. For player 2, it is always better to play the opposite strategy of player 1 (bold markings) since $A_2 \succ B_2$ in the left information set holds true and $B_2 \succ C_2$ in the right one. Player 1, anticipating this, sells their shares. In fact player 1 does not even need to anticipate, since the action of player 2 does not change the outcome for player 1; it is A_1 in either case.

The opposite case occurs if the informational edge fails to exceed the threshold value. This implies that the reactions are processed at the same time, resulting in a simultaneous game, which is depicted in Figure V–2b. The only change in outcome compared to the sequential game is in the event of both players selling. In the simultaneous game, there is an immediate oversupply and the associated price drop. Essentially, the leader from the sequential game cannot monetize their informational advantage. The result is that both players sell at a reduced price and/or the value of their remaining shares in the company is substantially diminished (C_i). Since both players have the same general preference structure, the game is also symmetric.

This incurs fundamental changes in the selection of strategies. The resulting simultaneous game is similar to a Game of Chicken (Rapoport & Chammah, 1966) in that there are two Nash equilibria in pure strategies when players choose opposing strategies (gray background). This is illustrated by the bolded best responses in Figure V–2b. Additionally, there is an equilibrium in mixed strategies, where the payoffs determine the probabilities of selling or keeping. The implication of the latter equilibrium is that, absent knowledge of the others' strategic decisions, players are likely to mix the strategies they act on. With an increasing number of players the probability that all (or most) sell, thereby inducing a crash, is significantly reduced. However, this is all under the assumption that players are aware of the change in the type of game.

3.2 The Influence of Information Technology on Game Form

This awareness is put into question if we consider how information technology has profoundly changed the way new information is received and processed. In the past, news was disseminated by, for instance, television, radio, telephone, or word of mouth, and processing used to be executed by human workers. Today, the central instrument for information dissemination is the Internet and processing has often been replaced by computer algorithms. The rise of the Internet and the cosmos of information technology surrounding it have revolutionized the way we acquire information and the speed at which we do so. This certainty has even spread beyond academia into popular culture. It culminates, for instance, in a famous essay by Carr (2008) where he describes the Internet as "the conduit for most of the information that flows through my eyes and ears and into my mind" (p. 90). As for financial markets in particular, technology has improved information dissemination, the quality of financial analysis, and the speed at which market participants communicate (Aldridge, 2010), resulting in a fundamental effect on game form determination. Since IT increases the speed of information dissemination and processing for most, perhaps even all, traders, it also substantially decreases the standard deviation of response times and, thereby, the informational edge potential leaders may hold over their followers. While information technology also accelerates market technologies (the threshold $\hat{\delta}$), there is no guarantee that the relative strength of both effects is equal. Hence, in some cases this acceleration effect may remove the first-mover advantage and turn a game that used to be sequential into a simultaneous one.

Furthermore, consider the influence of information technology on the variance of player technologies independent from acceleration. The large variety of channels used for the acquisition of information before the Internet, combined with the differing cognitive capabilities of the human workers involved, resulted in a much higher variance of response times when compared to today, even on a normalized level.

Nowadays, the Internet is the ultimate information channel and certain (automated) software packages dominate information processing. Therefore, we further argue that the dominance of the Internet and specific software packages causes a homogenization of response times across players (i.e. response times become more homogenous), diminishing the informational edge and increasing the likelihood of simultaneous games.

Hence, in some instances, information technology may change a game from sequential to simultaneous through the combined effect of acceleration and homogenization. The common knowledge assumption would require that players are aware of this change and always know the type of game they play. However, the common knowledge assumption is difficult to argue for, since it is more grounded in a desire for the mathematical tractability of games rather than an approximation of reality. Information acquisition, information technology, and information systems have long been associated with a competitive advantage, which is represented by the informational edge in the context of information-sensitive businesses. It is easy to argue that this mindset may still prevail even when companies are no longer creating edges but merely keeping pace. Furthermore, the vast majority of games will continue to be sequential, increasing the likelihood that simultaneity will not be anticipated. This causes two fundamental problems. First, from a leader's perspective, the sequential and simultaneous games are indistinguishable. Thus, players in a simultaneous game may believe themselves to be leaders in a sequential game simply because that is how such a situation has always presented itself—an argument that relates to game theoretic focal points (Schelling, 1990). In this case, players assume the type of game that seems natural, i.e. the sequential game. Nevertheless, even if players are aware of the general ambiguity, they may simply perceive themselves to be in the wrong game. Second, as shown above, the dominant strategy of the leader in the sequential game is selling, regardless of the subsequent action of the follower. This straightforward reaction to the event is easy to implement in a trading algorithm. The optimal strategy in a simultaneous game is less clear, since it depends on the actions of the other players. It requires a certain sentience of the trading algorithm, which substantially increases its complexity and may be infeasible.

To summarize, it is reasonable to assume that there are at least some instances where the advancement of information technology changes the form of the game. The rarer this occurrence is, the more likely it is that traders are not aware of this change. Even if they are, however, they may still employ the dominant strategy from the sequential game.

4 The Impact of Strategic Misperceptions

In this section, we analyze the effect of the strategic mismatch described in the previous paragraph using hypergame theory. We first investigate the simple two player scenario we have already introduced, followed by a generalization of our results to any given number of players. We close the section by discussing some numerical examples.

4.1 The Hypergame for Two Players

Recall Figure V–2a. The leader, in this case player 1, decides upon their action given the information from the event Θ. The follower can observe this action and select their own action given the information of Θ and the action by player 1. In the simultaneous game, all players react exclusively to Θ, as no other actions are observable. However, reacting exclusively to Θ is the leader's mark in the sequential game. Hence, if there is uncertainty about the game form, players may perceive the simultaneous game wrongly as a sequential game, in which they are the respective leader. To model this misperception, we use hypergame theory (Bennett, 1977, 1980). Sasaki & Kijima (2012) describe a hypergame as a collection of subjective games where "each agent believes that it is common knowledge among all the agents (who she thinks participate in the game) that the game they play is her own subjective game" (p. 723). More precisely, they define a hypergame $H = \left(I, (G^i)_{i \in I}\right)$ with a finite set of agents I and each agent's subjective game $G^i = \left(I^i, \mathrm{A}^i, \pi^i\right)$, where:

- I^i is the finite set of agents perceived by agent i, assuming $I^i \subseteq I$,

- $\mathrm{A}^i = \times_{j \in I^i} \mathrm{A}^i_j$ with A^i_j as the finite set of agent j's actions perceived by agent i,

- $\pi^i = \left(\pi^i_j\right)_{j \in I^i}$ with $\pi^i_j : \mathrm{A}^i \to \mathbb{R}$ as agent j's payoff as perceived by agent i.

Furthermore, they define $\alpha^* = \left(\alpha^*_i, \alpha^*_{-i}\right) \in \times_{i \in I} \mathrm{A}^i_i$ as a hyper Nash equilibrium of H iff $\alpha^*_i \in N_i(G^i)$ $\forall i \in I$, where $N_i(G^i)$ is the set of Nash actions of player i in their subjective game. An action $\alpha^*_i \in \mathrm{A}^i_i$ is player i's Nash action in G^i iff there exists $\alpha^i_{j \neq i} \in \mathrm{A}^i_{j \neq i}$, such that $\left(\alpha^*_i, \alpha^*_{j \neq i}\right) \in N(G^i)$ with $N(G^i)$ as the set of Nash equilibria of subjective game G^i. Thus, a hyper Nash equilibrium is defined as an outcome, where every player chooses an action that may result in a Nash equilibrium in the subjective game of that particular player. The set of hyper Nash equilibria of hypergame H is called $HN(H)$.

Adopting the definition by Sasaki & Kijima (2012), we derive a hypergame in which both players misperceive the simultaneous game as a sequential game. Their subjective games are illustrated in Figure V–3 and the hypergame is governed by the following equations (K = Keeping; S = Selling).

$$H = (I, (G^1, G^2)) \qquad\qquad (\text{V–1})$$

$$\text{with} \quad I = \{1, 2\} \qquad G^1 = (I, \mathrm{A}^1, \pi^1) \qquad G^2 = (I, \mathrm{A}^2, \pi^2)$$

$$\text{and} \quad \mathrm{A}^i_i = \{K, S\} \quad \mathrm{A}^i_{j \neq i} = \{KK, KS, SK, SS\} \quad \forall\ i, j \in I; j \neq i$$

$$\pi^i_i(\alpha^i_i = K) = B_i \qquad\qquad \pi^i_i(\alpha^i_i = S) = A_i$$
$$\pi^i_j(\alpha^i_j = KK) = B_j \qquad\qquad \pi^i_j(\alpha^i_j = KS \mid \alpha^i_i = K) = B_j$$

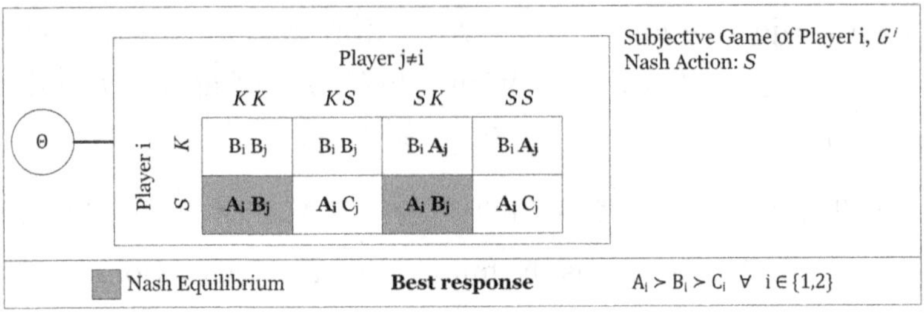

Figure V–3: Subjective game G^i

$$\pi_j^i(\alpha_j^i = KS \mid \alpha_i^i = S) = C_j \qquad \pi_j^i(\alpha_j^i = SK \mid \alpha_i^i = K) = A_j$$
$$\pi_j^i(\alpha_j^i = SK \mid \alpha_i^i = S) = B_j \qquad \pi_j^i(\alpha_j^i = SS \mid \alpha_i^i = K) = A_j$$
$$\pi_j^i(\alpha_j^i = SS \mid \alpha_i^i = S) = C_j \qquad A_n \succ B_n \succ C_n \quad \forall \ n \in I^i$$

Evidently, both subjective games are identical to the sequential game in Figure V–2a with the only difference being that each player sees themselves as leader in their own subjective game. Hence, the best responses and Nash equilibria reflect Figure V–2a, as well. Note particularly that $\pi_i^i(\alpha_i^i = S) > \pi_i^i(\alpha_i^i = K) \ \forall \ a_{j \neq i}^i \in A_{j \neq i}^i$. No matter what the other player does, player i always prefers to sell within their subjective game—selling is the dominant strategy and, thereby, the only Nash strategy of player i in the subjective game. Thus,

$$\pi_i^i(\alpha_i = S) > \pi_i^i(\alpha_i = K) \ \forall \ a_{j \neq i}^i \in A_{j \neq i}^i$$
$$\Rightarrow \ N_i(G^i) = \{S\} \ \forall \ i \in I. \tag{V-2}$$

If both players misperceive the simultaneous game as sequential, they consider themselves to be the leader of their respective subjective game. From the definition above, we derive a unique hyper Nash equilibrium as

$$N_i(G^i) = \{S\} \ \forall \ i \in I \ \Rightarrow \ HN(H) = \{(S, S)\}. \tag{V-3}$$

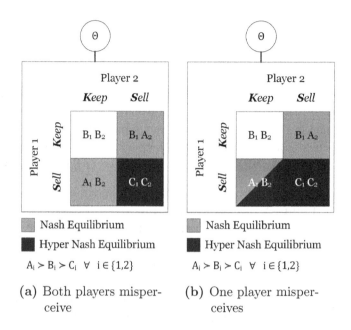

(a) Both players misper-
ceive

(b) One player misper-
ceives

Figure V–4: Sequential and simultaneous game

The implications of this result are illustrated in Figure V–4a. If both
players perceive a sequential game and themselves as leader, the true
game must be simultaneous. Otherwise, one player would go first and
the other player would realize that they are the follower. However,
the collective misperception causes the worst possible outcome in the
simultaneous game, since all players are selling and the price crashes.
Notice that in the objective game, this outcome does not constitute a
Nash equilibrium, as either player would improve their position by
deviating. Even when both players deviate, the outcome would still
be preferable. However, in the game as each player perceives it, selling
is the dominant strategy. Thus, we conclude the first central insight
of our model:

*If players misperceive the simultaneous game due to game form ambi-
guity, each player believes to be the leader within a sequential game.
Playing the leader's Nash strategy in the perceived sequential game in-*

duces a mutually detrimental outcome in the underlying simultaneous game.

Such an outcome can manifest, for instance, in a flash crash. Players react to news, but do not consider the simultaneous reaction of other players, thereby inducing an unwarranted price decline. This undervaluation of the asset subsequently causes other—or even the same—traders to buy shares. Prices increase to the appropriate level, resulting in the immediate rebound distinctive of flash crashes.

However, so far we have assumed that all players are misperceiving the game. Figure V–4b shows that implications become less clear once this assumption is dropped. In this particular case, player 1 misperceives the game and plays the sequential Nash strategy while player 2 correctly perceives the simultaneous nature of the game. From Figure V–4b, we can observe that both selling and keeping constitute Nash strategies for player 2. Hence, as evident in Figure V–4b, there are two Hyper Nash Equilibria in the game. Intuitively, this seems to soften the negative impact of player 1's misperception. We will further analyze the implications of only a subset of I misperceiving the game in the next subsection.

4.2 Misperceptions in Games with Many Players

In the model above, we analyzed a two player case and showed the detrimental effects of game misperception caused by game form ambiguity. However, in most financial markets, two player games are arguably rare. Yet if there are many players and only a small subset misperceives the game, this may already lead to suboptimal outcomes. To investigate this we need to revisit the sequential game and the simultaneous game to estimate the effect of misperceptions. For this, we redefine our set of players as

$$P = \{p_1, \ldots, p_i, \ldots, p_{\max}\}$$

where each player is defined by the tuple $p_i = (\alpha_i, t_i)$ $\forall$ $p_i \in P$. α_i is the action of player i and is defined as above as $\alpha_i \in \{S, K\}$, i.e. either selling or keeping. $t_i \in \{1, \ldots, t_{\max}\}$ is the position of player i in the game. A strictly sequential game is a game where all players have different positions, which implies that a player can observe the actions of all players with a smaller t-value. A strictly simultaneous game is a game where all players have the same t-value, i.e. no player moves first.

We further define M_τ as a subset of P with

$$M_\tau = \{p_m \in P \mid \alpha_m = S \wedge t_m \leq \tau\}.$$

Thus, M_τ contains those players, who sold before or at position τ. We can now generalize the payoff functions from the two-player game above as:

$$E(\pi_i \mid \alpha_i = K) = v,$$

$$E(\pi_i \mid \alpha_i = S \wedge t_i = \tau) = w_0 - w_1 |M_\tau|,$$

$$w_0 - w_1 > v > w_o - w_1 |P|.$$

The argument is similar to the one we made earlier in the paper. The payoff for keeping represents the long-term value of the stock, which is expected to be less than before the new negative information. If only a few people are selling, the price has not yet incorporated this negative information. Therefore, they get a payoff above v. However, if many players are selling, prices eventually drop below v. Note that the inequality above reflects the preference structure $A_i \succ B_i \succ C_i$ from the two-player game. For reasons of computability, we assume that the magnitudes of payoffs are symmetric among players. While asymmetric magnitudes may influence the probabilities in mixed strategies, they do not change the general implications of the following derivation.

In a strictly sequential game, every player knows their position and the actions of all players before. Hence, every player can calculate $E(\pi_i \mid \alpha_i = S \wedge t_i = \tau)$. If the resulting value exceeds v, it is a strictly dominant strategy for this player to sell, since payoffs do not depend on the actions of the following players. Similarly, if the value is less than v, the dominant strategy is to keep the shares. We assume that all players weakly prefer selling if the value is equal to v. It is important to note that it is always dominant for the player in the first position to sell, since $w_0 - w_1 > v$. Hence, the subgame perfect Nash Equilibrium is for the first t^* players to sell, where $w_0 - w_1 t^* \geq v > w_0 - w_1(t^* + 1)$, and for all other players to keep.

A strictly simultaneous game is governed by $t_i = 1 \; \forall \; p_i \in P$, i.e. all players are at the same position. We can calculate the Nash Equilibrium in mixed strategies by setting the expected payoff under each strategy equal, resulting in

$$v = w_0 - w_1 E(|M|). \tag{V--4}$$

$E(|M|)$ is the expected number of selling agents, which depends on the probability $q_i = \Pr(\alpha_i = S)$. Due to the strict symmetry in payoff functions, we know that $q_i = q \; \forall \; p_i \in P$. Thus, we can calculate the probability of agents selling from

$$v = w_0 - w_1 \sum_{k=1}^{|P|} \left[k \binom{|P| - 1}{k - 1} q^{k-1}(1 - q)^{|P|-k} \right]. \tag{V--5}$$

From this, we get $q^* = \frac{w_0 - (v + w_1)}{w_1(|P| - 1)}$ as the probability of selling in the mixed strategy Nash equilibrium. Naturally, the expected payoff for any player is v and the aggregate payoff is $|P|v$.

However, consider the case when player p_1 misperceives the game, such that they always play $\alpha_1 = S$. Their expected payoff is still v, since all other players mix their strategies according to the Nash equilibrium. Yet, the expected payoff to all other players changes since p_1 does not mix. More precisely the expected payoff is

$$E(\pi_{i \neq 1})$$
$$= (1 - q^*)v + q^* w_0$$
$$- q^* w_1 \left[2 + \sum_{k=1}^{|P|-2} \left[k \binom{|P| - 2}{k} (q^*)^k (1 - q^*)^{|P|-k-2} \right] \right]. \tag{V-6}$$

We show that this is always less than v in Section 9. However, this result is not surprising, since q^* represents the mixing probabilities in the unique mixed strategy Nash equilibrium of the game—a situation where no player has an incentive to unilaterally deviate. If player p_1 deviates and does not bear the burden of this deviation, the other players must. This implies that the aggregate payoffs of all players decreases.

We can further determine the impact of an increasing number of misperceiving players on payoffs by defining a set $R \subset P$, which contains all players that perceive themselves to be the sole leaders in a sequential game.

$$R = \{p_i \in P \mid E_i(M_1) = \{p_i\}\}$$

Based on this, we can adjust the payoff equation to describe the expected payoff of misperceiving ($p \in P \cup R$) and correctly perceiving traders ($p \in P \cap R$), respectively, with $\eta = |P| - |R|$.

$$E(\pi_i \mid p_i \in R \cup P)$$
$$= w_0 - w_1 \left(|R| + \sum_{k=1}^{\eta} \left[k \binom{\eta}{k} (q^*)^k (1 - q^*)^{\eta-k} \right] \right) \tag{V-7}$$

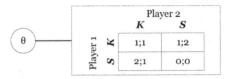

Figure V–5: Numerical example of two-player game

$$E(\pi_i \mid p_i \in R \cap P)$$
$$= (1 - q^*)v + q^* w_0 - q^* w_1 (|R| + 1)$$
$$- q^* w_1 \left(\sum_{k=1}^{\eta-1} \left[k \binom{\eta - 1}{k} (q^*)^k (1 - q^*)^{\eta - 1 - k} \right] \right) \tag{V–8}$$

The relationship between these payoffs naturally depends on the parameterization of w_0, w_1, v, and $|P|$. We will investigate this by using numerical examples in the next subsection.

4.3 Numerical Examples

We analyze the impact of an increasing number of players that perceive the incorrect game using numerical examples that conform to the payoff functions defined above. First, however, consider again the two-player case. We can satisfy the preference structure in Figure V–4 by setting $A = 2$, $B = 1$, $C = 0$ for each player. This also satisfies the condition from the previous subsection with $v = 1$, $w_0 = 4$, and $w_1 = 2$. The resulting simultaneous game is illustrated in Figure V–5.

Plugging the values into the equation above yields $q^* = 0.5$ as the probability of selling in the mixed strategy Nash equilibrium. Hence, the expected payoff in the simultaneous game would be 1 for either player. If player 1 misperceives the game, their payoff is still 1, while player 2's payoff decreases to 0.5, such that the aggregate payoff is 1.5. If both players misperceive, the aggregate payoff is zero. This corresponds with our finding that the aggregate payoff decreases with the number of misperceiving players.

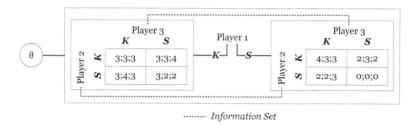

Figure V–6: Numerical example of three-player game

To further investigate this, we construct a three-player game with $v = 3$, $w_0 = 6$, and $w_1 = 2$, which is illustrated in Figure V–6. The information sets illustrate that the respective players are aware of the payoffs in the left and right panels but do not know which panel will be played. In this case, players mix with a probability of selling at $q^* = 0.25$. If there is no misperception, naturally the expected payoff is 3 for every player. Assuming that player 1 misperceives and plays S, their payoff is still 3. The payoffs for players 2 and 3 are $E(\pi_{i \neq 1}) = 2.625$. Thus, the aggregate payoff is 8.5 but the loss is borne by the players who have not misperceived.

However, considering the case of two players misperceiving shows that playing the subjective sequential game has detrimental effects on the misperceiving players, as well. If players 1 and 2 are misperceiving, the payoffs can be seen in the bottom row of the right panel of Figure V–6. Since player 3 is still mixing with $q^* = 0.25$, their payoff is 2.25, while the payoff for each misperceiving player is 1.5. The decrease in expected payoff is much more pronounced for players who misperceive than for the one who is not. The aggregate payoff is, thus, equal to 5.25.

We construct a final example, shown in Figure V–7, which uses the generalized payoff functions derived above (Equations V–7 and V–8) to illustrate a game with 25 players. The parameters are set to $v = 50$, $w_0 = 250$, and $w_1 = 10$. Naturally, the mixed-strategy Nash Equilibrium is at 50. As expected, an increasing number of misperceiving

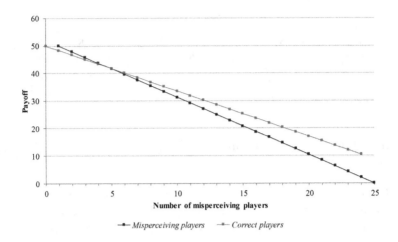

Figure V–7: Payoffs in 25-player game with increasing number of misperceptions

players decreases the expected payoff to correctly perceiving players at a constant rate. While a single misperceiving player receives a payoff of 50, an increase in $|R|$ causes a decrease of payoffs to those players, as well. The decrease also occurs at a constant rate and with a magnitude larger than that for correctly perceiving players. This results in misperceiving players eventually receiving lower payoffs than correctly perceiving players.

These examples show that even if only a small subset of agents misperceive the game that is objectively played, this has a detrimental impact on the aggregate performance. While these detriments are initially borne by the mixing players, misperceiving players suffer more as their number increases.

5 Implications, Interaction with Herding, and Limitations

In the past decades, information technology has continuously provided some traders in financial markets with competitive advantages—the ability to collect, process, and react to new information faster than their competitors. However, with an increasing number of traders adopting similar software and hardware, such advantages are more difficult to create than ever. Using a game theoretic model, we have shown that this can reshape the strategic environment of the market traders are operating in. While traders may perceive to be leaders in a sequential game, the homogeneity of technologies makes some games simultaneous.

Thereby, our model provides an intuitive explanation for flash crashes, i.e. sudden drops in market prices which quickly rebound once the game form ambiguity is resolved. Traders rely on the dominant sequential strategy—either because they are unaware of the change in game form or because the certain degree of awareness of other players required to determine the strategy in a simultaneous setting is too difficult to implement in an algorithm. As we have shown, this results in an unwarranted price drop. The model also explains the relatively few occurrences of flash crashes, since overall the vast majority of trades continues to occur in a sequential setting. Hence, a flash crash results from a disastrous combination of various simultaneous trades and other contributing factors.

As an example, it is easy to recognize how the misperception effect and herding (as a further contributing factor) amplify each other. Herd behavior is exhibited by uniformed traders who mirror the actions of players that are (perceived to be) better informed. Thereby, it amplifies the impact of actions by these informed traders. The resulting interaction between misperception and herding is illustrated in Figure V–8. It depicts a scenario in which an event causes the market price of a share to drop from 10 to 5. We have four observable time periods with $\tau = 0$ the period before the event and $\tau = 1$ the first position after the event. Any actions in a particular period are observable in the successive periods. Panel (a) illustrates the "traditional" sequential game—the fastest players are able to sell at prices above 5, which causes the price to adjust directly to 5 in period 1. Traders have no incentive to further sell any shares. Panel (b) depicts the simultaneous game with misperceptions. An increased propensity to sell (due to the perception of being the leader in a simultaneous game) causes the price to drop below the new equilibrial price in period 1, which is recovered in the following period. The effect of herding in a sequential game is shown in panel (c). Similar to the first scenario, the fast players realize their informational edge and the price adjusts to 5 in period 1. However, uninformed traders mirroring the actions of those fast players cause the price to drop further in period 2, which is recovered in the final period. Finally, panel (d) visualizes the combined effect of misperception in a simultaneous game and subsequent herd behavior. Some informed traders perceive themselves to be leaders in a sequential game and cause the price in period 1 to drop below the new equilibrium. In period 2, uninformed traders mirror the actions of those "false leaders" and exacerbate the price drop. Prices eventually adjust in period 3. Hence, uninformed traders follow the informed traders, unaware of the fact that the latter are essentially jumping off a cliff—similar to the popular myth that lemmings commit mass suicide during migration.

The strong interaction between misperceptions and herding as possible causes of flash crashes also makes it difficult to develop a cure for the problem. Consider again panel (d) in Figure V–8. While the price eventually adjusts to the equilibrial level in the final period, it is quite possible that the informed players, who caused the initial drop, benefit from this by acquiring the undervalued shares sold by the uninformed traders. This may well result in a net profit for the informed traders and a disincentive to change their strategies. Furthermore, we initially argued that such a strategic adaptation would be difficult to realize, because it requires an awareness of other players and their decision

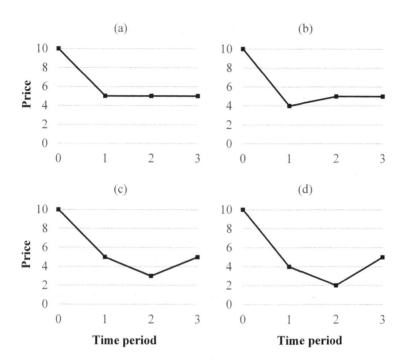

Figure V–8: Price movements in a sequential game (a), a simultaneous game with misperceptions (b), a simultaneous game with herding (c), and a simultaneous game with misperceptions and herding (d)

mechanisms. Therefore, another central implication of our model is that the occurrence of flash crashes in financial markets may be systemic and unavoidable given present market structures.

However, there are naturally various limitations to this study. The objective of this work was to provide a theoretical exploration of technological progress in IT as reasons for flash crashes—a theory that will need to be investigated empirically in future work. Furthermore, the assumptions associated with game-theoretic modelling always lead to a certain abstraction from reality. The sensitivity of the model to those assumptions needs to be analyzed. For instance, in the latter part of this paper, we have used the concept of a mixed-strategy Nash equilibrium as the benchmark case for our analysis. While there may be concerns that such an equilibrium is not necessarily realistic, the implications of our model stand despite this. Misperception leads to an overreliance on a particular strategic action (in our examples selling), which is likely to destabilize any market equilibrium. The mixed-strategy Nash equilibrium, therefore, only serves an illustrative purpose and was chosen due to its mainstream recognition. Nevertheless, further analyses into the effects of relaxing other assumptions of our model are necessary. In the next section, we demonstrate the broader relevance of our results using an application of the model to demand side management.

6 Application to Demand Side Management

Demand Side Management (DSM) as the automated control of energy demand is a new information-sensitive business sector that has evolved particularly during the past decade. It is necessitated by an increasing share of renewable energy sources in total energy generation, but also by a desire for increased energy efficiency. The fundamental value of information and the importance of behavioral factors in DSM have turned it into a prime research field for IS scholars (e.g. Watson et al., 2010; Strueker & Dinther, 2012; Feuerriegel et al., 2012).

On a residential level DSM has faced tremendous implementation challenges (Darby, 2006). Most appliances consume only small amounts of energy, at least when compared to industrial customers. Thus, DSM needs to be fully automated. Otherwise, the effort associated with having to manually adjust energy consumption easily outweighs possible financial benefits. An obvious solution would be to have these automated devices (e.g. washing machines; Katz et al., 2011) react to price signals. The energy retailer forwards the (adjusted) wholesale prices to the customers, thereby also forwarding information on current excess or lack of supply. However, this approach may create artificial demand peaks (Ramchurn et al., 2011), thereby exhibiting acceleration and homogenization effects.

Feuerriegel et al. (2013) argue that, while DSM can be profitable for retailers in general, higher information granularities are associated with disproportionately higher costs. This suggests that the intervals for price signals are quite large—minutes or even hours. These intervals represent $\hat{\delta}$—the threshold value incurred by the mechanism technology. It is an example how this parameter may not be limited by technological conditions but instead by economic considerations. DSM devices can react virtually instantaneously to these price signals—r_i is quite small. Also, since all the software has to do is receive the signal and decide whether to activate the device or not, response times are likely to be quite homogeneous. Hence, any informational edge that might emerge will be substantially smaller than $\hat{\delta}$, which results in a simultaneous game. The problem is that all DSM devices play leader strategies, since they exclusively react to the price signal. They make a binary decision between activating ($\alpha_i = 1$) and not activating ($\alpha_i = 0$) with the perceived payoff (per unit of energy) to device i, $\pi_i(\alpha_i)$, for price signal Θ and the mean (or expected future) price $\bar{\Theta}$ being:

$$\pi_i(1) = \bar{\Theta} - \Theta$$

$$\pi_i(0) = 0$$

Thus, for large downward price deviations, most devices will activate. However, taking the simultaneous game into account, their payoff is different and depends on how many devices activate—i.e. $|M|$:

$$\pi_i(\alpha_i, |M|) = \begin{cases} -C & \text{if } |M| \geq \xi \\ \bar{\Theta} - \Theta & \text{if} |M| < \xi \text{ and } \alpha_i = 1 \\ 0 & \text{otherwise} \end{cases} \qquad \text{(V–9)}$$

$$|M| = \sum_{i=1}^{|P|} \alpha_i$$

and C as a large positive real number as well as a threshold parameter ξ.

To explain the large costs incurred if ξ is exceeded, remember that a price drop (a small Θ) signals an oversupply of energy in the power grid. If responses are too homogeneous (exceeding ξ), demand increases enough to overcompensate the former oversupply. Gottwalt et al. (2011) refer to this phenomenon as an *avalanche effect* and it follows a similar rationale as flash crashes. As a result of the overcompensation, an artificial peak is created and additional supply reserves need to be activated. While this is expensive, the costs are not directly transferred to the customers (they would still receive $\bar{\Theta} - \Theta$ or 0, respectively), since they replied to the original price signal. However, the mechanism does not fulfill its purpose and due to the high costs of supply reserves, the retailer will eventually remove it. Thus, the cost C reflects the lost revenue to the customers from the DSM mechanism disappearing.

The DSM example shows that the insights gained from the model introduced in this paper not only apply to cases with a misperception of the game, but also for cases with a misconstruction of strategies for the game (although this misconstruction may reflect misperceptions by the constructor). It illustrates that it is important to consider how information technology accelerates different components of a mechanism (here, response times and signaling intervals). The model also illuminates possible solutions. On the one hand, the sequential game can be restored, such that the strategies match the game. For this $\hat{\delta}$ would need to be decreased enough for an informational edge to take effect—signaling intervals would need to be much shorter. However, as mentioned before, this approach is probably infeasible, since communication costs would explode. On the other hand, the strategies of the DSM devices could be adapted to reflect the simultaneous game. The devices would need to mix strategies. Yet, this is no trivial task, since the devices would need information about the number of other players in the system, which incurs privacy issues.

7 Conclusion

Over the past decades advances in information technology have fundamentally altered financial markets—a development that has only accelerated in recent years. Flash crashes, high-frequency trading, and the ongoing replacement of human actions by software algorithms are representative of this transition. However, the consequences of this process are not yet fully understood and an area of avid research. In this paper, we focus on one particular issue in this context by developing a theoretical model that relates information technology innovation to flash crashes. More specifically, we use hypergame theory to investigate how misperceptions of the strategic environment induced by this technological progress may be a factor contributing to flash crashes in financial markets. We first model how acceleration and homogenization of software and hardware used by traders introduce uncertainty concerning the form of the game. We subsequently analyze the effect of this ambiguity on strategy selection, game equilibria, and payoffs, as well as the interaction between game form misperceptions and herd behavior.

We find that misperceptions explain price movements that are characteristic for flash crashes very well—both the sudden price drop, as well as the subsequent rebound. Analyzing the interaction with herd behavior, we also note that traders whose strategic choice caused the crash may eventually benefit from it. This eliminates any incentives to change strategy selection and causes flash crashes to become unavoidable and a systemic challenge. Yet, the relevance of this research is not limited to the explanation of flash crashes, as we have outlined using the example of demand side management mechanisms. Flash crashes remain a rare occurrence—in fact, our model provides insights on why they are so rare. Furthermore, many exchanges have integrated trading curbs that cause a stop in trading following substantial price changes. However, these only address the symptoms, while the theoretical reasoning in this paper points to the underlying cause. Occasional game simultaneity may be a contributing factor to the general volatility of stock markets and other automated mechanisms, of which flash crashes and avalanche effects are just the most visible outcome. Hence, future research is needed on how further progress in information technology may require a rethinking of the fundamental design of these mechanisms. Our argument should also be considered in discussions on the potential limitations and implications of high-frequency trading.

8 References

Aldridge, I. (2010). *High-frequency trading: A practical guide to algorithmic strategies and trading systems.* Hoboken, NJ: Wiley.

Barlevy, G. & Veronesi, P. (2003). Rational panics and stock market crashes. *Journal of Economic Theory, 110*(2), 234–263.

Bennett, P. G. (1977). Toward a theory of hypergames. *Omega, 5*(6), 749–751.

Bennett, P. G. (1980). Hypergames: Developing a model of conflict. *Futures, 12*(6), 489–507.

Budish, E. B., Cramton, P. & Shim, J. J. (2013). The High-Frequency Trading Arms Race: Frequent Batch Auctions as a Market Design Response. *SSRN Working Paper Series.*

Carr, N. G. (2004). *Does IT matter? Information technology and the corrosion of competitive advantage.* Boston, MA: Harvard Business School Press.

Carr, N. G. (2008). Is Google Making Us Stupid? *Yearbook of the National Society for the Study of Education*, *107*(2), 89–94.

CFTC and SEC. (2010). *Findings Regarding the Market Events of May 6, 2010. Report of the Staffs of the CFTC and SEC to the Joint Advisory Committee on Emerging Regulatory Issues.* Washington, DC.

Chlistalla, M. (2011). *High-frequency trading: Better than its reputation?*

Clemons, E. K. & Weber, B. W. (1990). London's Big Bang: A Case Study of Information Technology, Competitive Impact, and Organizational Change. *Journal of Management Information Systems*, *6*(4), 41–60.

Copeland, T. E. & Friedman, D. (1987). The Effect of Sequential Information Arrival on Asset Prices: An Experimental Study. *The Journal of Finance*, *42*(3), 763–797.

Darby, S. (2006). *The Effectivesness of Feedback on Energy Consumption.*

Devenow, A. & Welch, I. (1996). Rational herding in financial economics. *European Economic Review*, *40*(3-5), 603–615.

Easley, D., Lopez de Prado, M. M. & O'Hara, M. (2012). Flow Toxicity and Liquidity in a High-frequency World. *Review of Financial Studies*, *25*(5), 1457–1493.

Easley, D., Lpez de Prado, Marcos M & O'Hara, M. (2011). The Microstructure of the "Flash Crash": Flow Toxicity, Liquidity Crashes, and the Probability of Informed Trading. *The Journal of Portfolio Management*, *37*(2), 118–128.

Feuerriegel, S., Bodenbenner, P. & Neumann, D. (2013). Is More Information Better Than Less? Understanding The Impact Of Demand Response Mechanisms In Energy Markets. *ECIS 2013 Completed Research*, Paper 167.

Feuerriegel, S., Strüker, J. & Neumann, D. (2012). Reducing Price Uncertainty through Demand Side Management. *ICIS 2012 Proceedings*, Paper 7.

Filimonov, V. & Sornette, D. (2012). Quantifying reflexivity in financial markets: Toward a prediction of flash crashes. *Physical Review E*, *85*(5), 056108.

Gomber, P. & Haferkorn, M. (2013). High-Frequency-Trading - High-Frequency-Trading Technologies and Their Implications for Electronic Securities Trading. *Business & Information Systems Engineering*, *5*(2), 97–99.

Gottwalt, S., Ketter, W., Block, C., Collins, J. & Weinhardt, C. (2011). Demand side management—A simulation of household behavior under variable prices. *Energy Policy*, *39*(12), 8163–8174.

Groth, S. (2010). Enhancing Automated Trading Engines To Cope With News-Related Liquidity Shocks. *ECIS 2010 Proceedings*, Paper 111.

Katz, R. H., Culler, D. E., Sanders, S., Alspaugh, S., Chen, Y., Dawson-Haggerty, S., ... Shankar, S. (2011). An information-centric energy infrastructure: The Berkeley view. *Sustainable Computing: Informatics and Systems*, *1*(1), 7–22.

Kirilenko, A. A., Kyle, A. S., Samadi, M. & Tuzun, T. (2011). The Flash Crash: The Impact of High Frequency Trading on an Electronic Market. *SSRN Working Paper Series*.

Lattemann, C., Loos, P., Gomolka, J., Burghof, H.-P., Breuer, A., Gomber, P., ... Zajonz, R. (2012). High Frequency Trading - Costs and Benefits in Securities Trading and its Necessity of Regulations. *Business & Information Systems Engineering*, *4*(2), 93–108.

Lee, H. G. & Clark, T. H. (1996/1997). Market Process Reengineering through Electronic Market Systems: Opportunities and

Challenges. *Journal of Management Information Systems*, *13*(3), 113–136.

Lucas, H., Agarwal, R., Sawy, O. E. & Weber, B. (2013). Impactful Research on Transformational Information Technology: An Opportunity to Inform New Audiences. *Management Information Systems Quarterly*, *37*(2), 371–382.

Lux, T. (1995). Herd Behaviour, Bubbles and Crashes. *The Economic Journal*, *105*(431), 881–896.

Premkumar, G., Ramamurthy, K. & Saunders, C. S. (2005). Information Processing View of Organizations: An Exploratory Examination of Fit in the Context of Interorganizational Relationships. *Journal of Management Information Systems*, *22*(1), 257–298.

Ramchurn, S. D., Vytelingum, P., Rogers, A. & Jennings, N. (2011). Agent-based control for decentralised demand side management in the smart grid. In *The 10th International Conference on Autonomous Agents and Multiagent Systems* (pp. 5–12). Ann Arbor: IFAAMAS.

Rapoport, A. & Chammah, A. M. (1966). The Game of Chicken. *American Behavioral Scientist*, *10*(3), 10–28.

Sakaki, T., Okazaki, M. & Matsuo, Y. (2010). Earthquake shakes Twitter users: real-time event detection by social sensors. *WWW '10 Proceedings*, 851–860.

Sasaki, Y. & Kijima, K. (2012). Hypergames and bayesian games: A theoretical comparison of the models of games with incomplete information. *Journal of Systems Science and Complexity*, *25*(4), 720–735.

Schelling, T. (1990). *The Strategy of Conflict (Reprint)*. Cambridge, MA: Harvard University Press.

Strueker, J. & Dinther, C. (2012). Demand Response in Smart Grids: Research Opportunities for the IS Discipline. *AMCIS 2012 Proceedings*, Paper 7.

Walczak, S. (1999). Gaining Competitive Advantage for Trading in Emerging Capital Markets with Neural Networks. *Journal of Management Information Systems, 16*(2), 177–192.

Wang, E., Tai, C.-F. & Grover, V. (2013). Examining the Relational Benefits of Improved Interfirm Information Processing Capability in Buyer–Supplier Dyads. *Management Information Systems Quarterly, 37*(1), 149–173.

Watson, R., Boudreau, M.-C. & Chen, A. (2010). Information Systems and Environmentally Sustainable Development: Energy Informatics and New Directions for the IS Community. *Management Information Systems Quarterly, 34*(1), 23–38.

White, M. (2011). Information anywhere, any when: The role of the smartphone. *Business Information Review, 27*(4), 242–247.

Xu, S. & Zhang, X. (2009). How Do Social Media Shape the Information Environment in the Financial Market? *ICIS 2009 Proceedings*, Paper 56.

Zhang, S. & Riordan, R. (2011). Technology and Market Quality: The Case of High Frequency Trading. *ECIS 2011 Proceedings*, Paper 95.

9 Appendix

We show that $E(\pi_{i \neq 1})$ is always less than v, which implies that a single player perceiving the wrong game form in a game with $|P|$ players reduces the payoffs for all other players, albeit not their own. Recall that $w_0 - w_1 > v > w_0 - w_1|P|$, which can only be satisfied if $w_1 > 0$. Taking this for $q^* = \frac{w_o - (v + w_1)}{w_1(|P| - 1)}$ into account, q^* is also larger than zero (since $w_0 - (v + w_1) > 0 \iff w_0 - w_1 > v$) and less than one (since $w_0 - (v + w_1) < w_1(|P| - 1) \iff v > w_0 - w_1|P|$). This confirms that q^* satisfies the assumptions for a probability. From the equation governing the mixed strategy Nash equilibrium of the simultaneous game, we know that

$$v = w_0 - w_1 \sum_{k=1}^{|P|} \left[k \binom{|P| - 1}{k - 1} q^{k-1}(1 - q)^{|P|-k} \right].$$

Hence,

$$E(\pi_{i \neq 1}) < v$$

$$\iff E(\pi_{i \neq 1}) < w_0 - w_1 \sum_{i=k}^{|P|} \left[k \binom{|P| - 1}{k - 1} q^{k-1}(1 - q)^{|P|-k} \right]$$

with

$$E(\pi_{i \neq 1}) = (1 - q^*)v + q^* w_0$$

$$- q^* w_1 \left[2 + \sum_{k=1}^{|P|-2} \left[k \binom{|P| - 2}{k} (q^*)^k (1 - q^*)^{|P|-k-2} \right] \right].$$

Replacing v in the equation for $E(\pi_{i \neq 1})$ and expanding the sums and binomials on each side of the inequality yields

$$E(\pi_{i \neq 1}) < v$$
$$\iff w_0 + w_1(q^*(q^* - |P|) - 1) < w_0 + w_1(q^*(1 - |P|) - 1)$$
$$\iff q^*(q^* - |P|) - 1 < q^*(1 - |P|) - 1$$
$$\iff q^* < 1$$

As shown above, this is always the case.

Printed in the United States
By Bookmasters